Applied Mathematical Sciences | Volume 52

Applied Mathematical Sciences

1. John: **Partial Differential Equations**, 4th ed.
2. Sirovich: **Techniques of Asymptotic Analysis.**
3. Hale: **Theory of Functional Differential Equations**, 2nd ed.
4. Percus: **Combinatorial Methods.**
5. von Mises/Friedrichs: **Fluid Dynamics.**
6. Freiberger/Grenander: **A Short Course in Computational Probability and Statistics.**
7. Pipkin: **Lectures on Viscoelasticity Theory.**
8. Giacaglia: **Perturbation Methods in Non-Linear Systems.**
9. Friedrichs: **Spectral Theory of Operators in Hilbert Space.**
10. Stroud: **Numerical Quadrature and Solution of Ordinary Differential Equations.**
11. Wolovich: **Linear Multivariable Systems.**
12. Berkovitz: **Optimal Control Theory.**
13. Bluman/Cole: **Similarity Methods for Differential Equations.**
14. Yoshizawa: **Stability Theory and the Existence of Periodic Solutions and Almost Periodic Solutions.**
15. Braun: **Differential Equations and Their Applications**, 3rd ed.
16. Lefschetz: **Applications of Algebraic Topology.**
17. Collatz/Wetterling: **Optimization Problems.**
18. Grenander: **Pattern Synthesis: Lectures in Pattern Theory, Vol I.**
19. Marsden/McCracken: **The Hopf Bifurcation and its Applications.**
20. Driver: **Ordinary and Delay Differential Equations.**
21. Courant/Friedrichs: **Supersonic Flow and Shock Waves.**
22. Rouche/Habets/Laloy: **Stability Theory by Liapunov's Direct Method.**
23. Lamperti: **Stochastic Processes: A Survey of the Mathematical Theory.**
24. Grenander: **Pattern Analysis: Lectures in Pattern Theory, Vol. II.**
25. Davies: **Integral Transforms and Their Applications.**
26. Kushner/Clark: **Stochastic Approximation Methods for Constrained and Unconstrained Systems.**
27. de Boor: **A Practical Guide to Splines.**
28. Keilson: **Markov Chain Models—Rarity and Exponentiality.**
29. de Veubeke: **A Course in Elasticity.**
30. Sniatycki: **Geometric Quantization and Quantum Mechanics.**
31. Reid: **Sturmian Theory for Ordinary Differential Equations.**
32. Meis/Markowitz: **Numerical Solution of Partial Differential Equations.**
33. Grenander: **Regular Structures: Lectures in Pattern Theory, Vol. III.**
34. Kevorkian/Cole: **Perturbation Methods in Applied Mathematics.**
35. Carr: **Applications of Centre Manifold Theory.**

(continued after Index)

M. Chipot

Variational Inequalities and Flow in Porous Media

Springer-Verlag
New York Berlin Heidelberg Tokyo

M. Chipot
Department of Mathematics
University of Nancy I
B.P. 239-54506 Vandoeuvre
Cedex
France

AMS Classification: 76S05, 49A29

Library of Congress Cataloging in Publication Data
Chipot, M. (Michel)
   Variational inequalities and flow in porous media.
   (Applied mathematical sciences ; v. 52)
   Bibliography: p.
   Includes index.
   1. Fluid dynamics.   2. Porous materials.
3. Variational inequalities (Mathematics)   I. Title.
II. Series: Applied mathematical sciences
(Springer-Verlag New York Inc.) ; v. 52.
QA1.A647   vol. 52   [QA911]   510s   [532'.051]   84-5598

With 13 Illustrations

Printed and bound by R.R. Donnelley & Sons, Harrisonburg, Virginia.
Printed in the United States of America.

9 8 7 6 5 4 3 2 1

ISBN 0-387-96002-3 Springer-Verlag New York Berlin Heidelberg Tokyo
ISBN 3-540-96002-3 Springer-Verlag Berlin Heidelberg New York Tokyo

Preface

These notes are the contents of a one semester graduate course which I taught at Brown University during the academic year 1981-1982. They are mainly concerned with regularity theory for obstacle problems, and with the dam problem, which, in the rectangular case, is one of the most interesting applications of Variational Inequalities with an obstacle.

Very little background is needed to read these notes. The main results of functional analysis which are used here are recalled in the text. The goal of the two first chapters is to introduce the notion of Variational Inequality and give some applications from physical mathematics. The third chapter is concerned with a regularity theory for the obstacle problems. These problems have now invaded a large domain of applied mathematics including optimal control theory and mechanics, and a collection of regularity results available seems to be timely. Roughly speaking, for elliptic variational inequalities of second order we prove that the solution has as much regularity as the obstacle(s). We combine here the theory for one or two obstacles in a unified way, and one of our hopes is that the reader will enjoy the wide diversity of techniques used in this approach.

The fourth chapter is concerned with the dam problem. This problem has been intensively studied during the past decade (see the books of Baiocchi-Capelo and Kinderlehrer-Stampacchia in the references). The relationship with Variational Inequalities has already been quoted above. Starting with a new point of view introduced by Brezis-Kinderlehrer-Stampacchia, and in a different setting by Alt, we develop the theory for general domains, and with relatively elementary techniques we give the main results on the subject including a first study of the free boundary (i.e., the one limiting the wet set of the porous medium considered).

I am especially indebted to H. Brezis from whom I learned most of this subject.

I wish also to express my thanks to my colleagues in the Applied Mathematics Division of Brown University and particularly to Constantine Dafermos and Jack Hale who helped to create such a friendly atmosphere in the Division.

Thomas Sideris and Jalal Shatah helped me to correct the manuscript; I thank them very much for their friendly assistance.

Finally, I thank Roberta Weller and Katherine MacDougall for the excellent typing of the manuscript.

Michel Chipot
Providence, R.I.
February 1982

Table of Contents

Page

PREFACE   v

CHAPTER 1. ABSTRACT EXISTENCE AND UNIQUENESS RESULTS FOR
            SOLUTIONS OF VARIATIONAL INEQUALITIES   1

        1.1. Fixed Point Theorems   1
        1.2. Motivation   1
        1.3. Existence and Uniqueness Results   3
              Comments   9

CHAPTER 2. EXAMPLES AND APPLICATIONS   10

        2.1. Some Functional Analysis   10
        2.2. The Dirichlet Problem   14
        2.3. The Obstacle Problem   16
        2.4. Elastic Plastic Torsion Problems   17
        2.5. Nonlinear Operators   18
        2.6. Fourth Order Variational Inequalities   20
              Comments   21

CHAPTER 3. THE OBSTACLE PROBLEMS: A REGULARITY THEORY   22

        3.1. Monotonicity Results   22
        3.2. The Penalty Method   24
        3.3. $W^{2,p}(\Omega)$-Regularity $(2 \leq p < +\infty)$ 27
        3.4. Some Complementary Results   30
        3.5. $W^{2,\infty}(\Omega)$-Regularity   33
        3.6. $W^{1,\infty}(\Omega)$-Regularity   50
        3.7. $W^{1,p}(\Omega)$-Regularity $(2 < p < +\infty)$ 60

        Appendix. L^p-Estimates for the Solution of the
                   Dirichlet Problem   67
                   Comments   73

CHAPTER 4. THE DAM PROBLEM   74

        4.1. Statement of the Problem   74
        4.2. Some Properties of (p,χ) Solution of (P)   82
        4.3. S_3-Connected Solutions   90
        4.4. Uniqueness of S_3-Connected Solutions   101
        4.5. Some Monotonicity Results for the Free Boundary   106
              Comments   110

REFERENCES   111

INDEX   117

Chapter 1

Abstract Existence and Uniqueness Results for Solutions of Variational Inequalities

1.1. Fixed Point Theorems

In the different sections, we shall use repeatedly the following well known theorem:

Theorem 1.1: (Schauder) Let K be a compact convex subset of a Banach space X. If $F : K \to K$ is continuous, then F has a fixed point in K.

For a proof see [65] or [67].

As an easy corollary one can prove

Corollary 1.2. Let K be a closed and bounded convex subset of a Banach space X and $F : K \to K$ a completely continuous mapping. Then F has a fixed point in K.

Proof: Let K' = the closed convex hull of $F(K)$. It follows from Mazur's Theorem that K' is compact and thus F mapping K' into K' has a fixed point in K'. (See also [65].)

1.2. Motivation

Let X be, for instance, a Banach space and $f : X \to \mathbb{R}$ be a differentiable map. The problem of finding u such that

$$u \in X, \quad f(u) \le f(v) \quad \forall v \in X \tag{1.1}$$

leads us to consider the equation in X' (the topological dual of X)

$$f'(u) = 0. \tag{1.2}$$

Now if we consider the more general problem of finding a solution u of

$$u \in K, \quad f(u) \le f(v) \quad \forall v \in K \tag{1.3}$$

where K is a proper convex subset of X, then (1.2) may fail. Indeed
choose K = [0,1] and f as in the figure below:

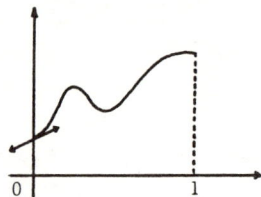

0 is a point where f achieves its minimum, but f'(0) ≠ 0. (Note also
the importance of considering closed sets K in problems like (1.3) since
on (0,1), f doesn't achieve its minimum.)

For problem (1.3), although a solution u may not satisfy (1.2),
that is to say:

$$\langle f'(u),v\rangle = 0 \qquad \forall v \in K \tag{1.4}$$

($\langle\ ,\ \rangle$ denoting the duality between X',X), we can prove that u satisfies:

$$u \in K, \quad \langle f'(u),v - u\rangle \geq 0 \qquad \forall v \in K. \tag{1.5}$$

Indeed, for all $v \in K$, let

$$\Phi(t) = f(u + t(v - u)) \qquad \forall t \in [0,1].$$

(Note that $u + t(v - u) \in K$ for $t \in [0,1]$, $u,v \in K$ and since f is
differentiable, we have $\Phi(t) - \Phi(0) = t \langle f'(u), v - u\rangle + o(t)$.) Now
from (1.3), we deduce

$$\Phi(t) \geq \Phi(0) \qquad \forall t \in [0,1].$$

Hence,

$$0 \leq \lim_{t\to 0} \frac{\Phi(t) - \Phi(0)}{t} = \langle f'(u),v - u\rangle \qquad \forall v \in K.$$

We will call (1.5) a Variational Inequality (V.I.). Moreover, if we also
assume that f is convex, then (1.5) characterizes exactly the points of
K for which (1.3) holds. Indeed, in this case we have (since $\Phi(t)$ is
also convex)

$$\Phi(1) \geq \Phi(0) + \Phi'(0)$$

$$\iff \quad f(v) \geq f(u) + \langle f'(u),v - u\rangle \geq f(u) \qquad \forall v \in K.$$

We can summarize the above result in

Proposition 1.3. Let $f : X \to \mathbb{R}$ be convex and differentiable and K a closed convex set of X, then

$$u \in K, \quad f(u) \leq f(v) \qquad \forall v \in K$$

is equivalent to

$$u \in K, \quad \langle f'(u), v - u \rangle \geq 0 \quad \forall v \in K.$$

Thus V.I.'s arise naturally in variational problems on convex sets. In the next paragraph, we shall investigate more closely problems like (1.5). More precisely, let f be an element of X', $A : K \to X'$. We shall study problems of the type:

$$u \in K, \quad \langle Au, v - u \rangle \geq \langle f, v - u \rangle \quad \forall v \in K. \tag{1.6}$$

Before solving (1.6), let us quote a special case where Proposition 1.3 arises, and moreover, where we know that a minimum is achieved.

Let K be a nonempty closed convex set in a Hilbert space H with the inner product $(\; , \;)$ and let f be an element of H. Then it is well known that there exists a unique u in K, called the projection of f on K, such that with $||v|| = (v,v)^{1/2}$ we have

$$u \in K, \quad ||u-f||^2 \leq ||v-f||^2 \quad \forall v \in K.$$

Moreover, u is the unique element such that (apply, for instance, Proposition 1.3)

$$u \in K, \quad (u,v-u) \geq (f,v-u) \quad \forall v \in K \tag{1.7}$$

and thus projection on a closed convex set in Hilbert spaces provide us with a very large class of V.I. Note also that in this case uniqueness holds (see corollary 1.13). In the sequel, we shall denote by P_K the map $f \to u$ where u is the solution of (1.7).

1.3. Existence and Uniqueness Results

1.3.1. The finite dimensional case.

Let X be a finite dimensional space, X' its dual, and $\langle , \rangle$ the pairing between X',X.

Theorem 1.4. Let K be a nonempty compact convex subset of X and A a continuous mapping of K into X'. Then for every $f \in X'$ there exists a solution u of the problem:

$$u \in K, \quad \langle Au, v - u \rangle \geq \langle f, v - u \rangle \quad \forall v \in K. \tag{1.8}$$

<u>Proof</u>: Choose a Euclidean structure on X i.e., an inner product $(\ ,\ )$
and denote by $j : X' \to X$ the mapping defined by

$$<x',x> = (j(x'),x) \qquad \forall x \in X. \tag{1.8}$$

If P_K denotes the projection of X on K for the above Euclidean
structure, then the mapping

$$x \longmapsto P_K(x - j(Ax - f))$$

is continuous from K into K and has a fixed point u which satisfies
(see 1.7))

$$u \in K, (u,v - u) \geq (u - j(Au - f),v - u) \qquad \forall v \in K,$$

and so by (1.8) u satisfies (1.6).

<u>Remark 1.5</u>. Note that in the case where K is unbounded (1.6) doesn't
necessarily have a solution. Indeed, choose for example $X = X' = \mathbb{R} = K$
with the pairing

$$<x',x> = x'x .$$

For $A : \mathbb{R} \to (0,+\infty)$, the problem

$$Au \cdot (v - u) \geq 0 \qquad \forall v \in \mathbb{R}$$

is equivalent to Au = 0, which doesn't have a solution. Thus, in the case
where K is unbounded some assumptions must be added.

To study the case when K is unbounded, we make the following definition:

<u>Definition 1.6</u>. Let K be a convex unbounded subset of a Banach space X
with dual space X'. We shall say that $A : K \to X'$ is coercive on K
if there exists $v_0 \in K$ such that

$$<Av - Av_0, v - v_0>/||v-v_0|| \to +\infty \text{ whenever } v \in K, ||v|| \to +\infty \tag{1.9}$$

$(< , >$ is the pairing between $X',X, ||\ ||$ the norm on X).

With this definition we can now prove

<u>Theorem 1.7</u>. Let K be a closed convex subset of X and $A : K \to X'$ a
continuous coercive mapping. Then for every f in X' there exists a
solution u of

$$u \in K, <Au,v - u> \geq <f,v - u> \quad \forall v \in K. \tag{1.10}$$

<u>Proof</u>: Since the map $x \longmapsto Ax - f$ is continuous and coercive, if we
can solve (1.6) in the case f = 0 then general case will follow. Thus
assume f = 0, and let B_R denote the closed ball of center 0 and radius

R in X (for the norm $||\ ||$ in X), u_R the solution of

$$u_R \in K \cap B_R, \quad <Au_R, v - u_R> \geq 0 \quad \forall v \in K \cap B_R. \tag{1.10}$$

Choose $R > ||v_0||$ with v_0 as in (1.9). Then we have by (1.10)

$$<Au_R, v_0 - u_R> \geq 0. \tag{1.11}$$

Moreover,

$$<Au_R, v_0 - u_R> = -<Au_R - Av_0, u_R - v_0> + <Av_0, v_0 - u_R>$$

$$\leq -<Au_R - Av_0, u_R - v_0> + ||Av_0||' \cdot ||u_R - v_0||$$

$$\leq ||u_R - v_0||(-<Au_R - Av_0, u_R - v_0>/||u_R - v_0|| + ||Av_0||')$$

where $||\ ||'$ denotes the strong dual norm in X'. Now if $||u_R|| = R$
for all R we may choose R big enough so that the above inequality and
coercivesness of A imply

$$<Au_R, v_0 - u_R> < 0$$

which contradicts (1.11). So there exists an R such that $||u_R|| < R$.
Now for every v on K we can choose $\varepsilon > 0$ small enough such that
$u_R + \varepsilon(v - u_R) \in K \cap B_R$ and thus by (1.10)

$$<Au_R, \varepsilon(v - u_R)> \geq 0 \quad \forall v \in K$$

$$\longleftrightarrow \quad <Au_R, v - u_R> \geq 0 \quad \forall v \in K$$

which proves that u_R is the solution of (1.6).

1.3.2. The infinite dimensional case.

In this part we shall denote by X a reflexive Banach space, X' its
dual, <,> the pairing between X' and X, K a closed convex set of X.
A will be a mapping of K into X' and $||\ ||$ will denote the norm
in X.

Let us give some definitions.

Definition 1.8. We shall say that A is monotone if

$$<Av - Au, v - u> \geq 0 \quad \forall v, u \in K \tag{1.12}$$

and A is strictly monotone if equality holds in (1.12) only for $v = u$.

Definition 1.9. We shall say that A is continuous on finite dimensional
subspaces if for every finite dimensional subspaces M of X the mapping
$A : K \cap M \to X'$ is weakly continuous, that is to say, if for all $x \in X$,

$v \longmapsto \langle Av,x \rangle$ is continuous on $K \cap M$.

With these definitions we now can state

Theorem 1.10. Let K be a nonempty closed convex subset of X, a reflexive Banach space. Let $f \in X'$ and $A : K \to X'$ be a monotone map continuous on finite dimensional subspaces. Then, if

   (i) K is bounded, or
   (ii) A is coercive on K
there exists a solution u of the V.I.

$$u \in K, \quad \langle Au, v - u \rangle \geq \langle f,v - u \rangle \quad \forall v \in K. \qquad (1.6)$$

Moreover, if A is strictly monotone, the solution of (1.6) is unique. Before giving a proof of this theorem, let us note the following useful lemma.

Minty's Lemma. Let K be a nonempty closed convex subset of a Banach space X and A a monotone operator from K into X' which is continuous on finite dimensional subspaces of X. Then for $f \in X'$

$$u \in K, \langle Au,v - u \rangle \geq \langle f,v - u \rangle \quad \forall v \in K \qquad (1.6)$$

$\Longleftrightarrow$

$$u \in K, \langle Av,v - u \rangle \geq \langle f,v - u \rangle \quad \forall v \in K. \qquad (1.13)$$

Proof: For ($\Rightarrow$) it suffices to remark that by monotonicity

$$\langle Av,v - u \rangle \geq \langle Au,v - u \rangle \geq \langle f,v - u \rangle \quad \forall v \in K.$$

For ($\Leftarrow$) replace v in (1.13) with $u + t(v - u)$ for $t \in (0,1]$. Then we have

$$\langle A(u + t(v - u)),t(v - u) \rangle \geq \langle f,t(v - u) \rangle \quad \forall v \in K.$$

But t is strictly positive, hence

$$\langle A(u + t(v - u)),v - u \rangle \geq \langle f,v - u \rangle \quad \forall v \in K.$$

Now using the continuity of A on finite dimensional subspaces and letting t go to zero, we obtain (1.6).

We are now able to prove Theorem 1.10.

Proof of Theorem 1.10: Consider case (i). Without loss of generality we assume $f = 0$.

Consider

$$C(v) = \{u \in K | \langle Av,v - u \rangle \geq 0\}.$$

It is easy to see that $C(v)$ is a weakly closed subset of K which is bounded. So $C(v)$ is weakly compact, and if

$$\bigcap_{v \in K} C(v) = \emptyset$$

we can find $v_1, \ldots, v_n$ in K such that

$$C(v_1) \cap C(v_2) \cap \ldots \cap C(v_n) = \emptyset. \tag{1.14}$$

But now consider M the subspace of X spanned by $v_1 \ldots v_n$. Denote by i the canonical injection of $M \to X$ and by r the "restriction" mapping defined by

$$r(x') = x'|_M \qquad \forall x' \in X'.$$

By applying the result of Theorem 1.4 and Minty's Lemma to the operator

$$v \to r \bullet A \bullet i(v)$$

we have that there exists a solution of the problem

$$u \in K \cap M, \quad \langle Av, v - u \rangle \geq 0 \qquad \forall v \in K \cap M.$$

Of course such a u belongs to $C(v_1) \cap \ldots \cap C(v_n)$ which gives a contradiction to (1.14), and thus $\bigcap_{v \in K} C(v) \neq \emptyset$. The theorem in case (i), now follows from another application of Minty's Lemma. For (ii) the proof now follows easily as in Theorem 1.7.

To prove uniqueness in the case of strictly monotone operators, it suffices to note that if u_1, u_2 are two solutions of (1.6), then we have

$$u_1 \in K, \quad \langle Au_1, v - u_1 \rangle \geq \langle f, v - u_1 \rangle \qquad \forall v \in K$$

$$u_2 \in K, \quad \langle Au_2, v - u_2 \rangle \geq \langle f, v - u_2 \rangle \qquad \forall v \in K.$$

Thus setting $v = u_2$ in the first inequality and $v = u_1$ in the second and adding leads to

$$\langle Au_1 - Au_2, u_1 - u_2 \rangle \leq 0.$$

By strict monotonicity this gives $u_1 = u_2$ and the result.

As a very useful corollary, let us note (compare with (1.2)):

<u>Corollary 1.11</u>. Let X be a reflexive Banach space, X' its dual. Let A be a monotone, coercive operator of X into X' which is continuous on finite dimensional subspaces of X. Then A is onto i.e., for all $f \in X'$ one can solve the equation:

$$Au = f. \tag{1.15}$$

If A is strictly monotone, the solution of (1.15) is unique.

Proof: By Theorem 1.10, we deduce that there exists u   such that

$$u \in X, \quad \langle Au, v - u \rangle \geq \langle f, v - u \rangle \quad \forall v \in X.$$

Taking $v = u \pm w$, $w \in X$ we have

$$\langle Au, w \rangle = \langle f, w \rangle \quad \forall w \in X$$

and thus (1.15) and the theorem follows.

Remark 1.12. In view of this corollary, the coerciveness assumption seems more natural. Indeed, if A is a monotone mapping from $\mathbb{R}$ into $\mathbb{R}$, it is clear that for A to be onto one needs that

$$|Ax| \to +\infty \quad \text{when} \quad |x| \to +\infty$$

i.e.,

$$\frac{|Ax||x|}{|x|} = \frac{Ax \cdot x}{|x|} \to +\infty \quad \text{when} \quad |x| \to +\infty.$$

In the particular case of Hilbert spaces, let us note also:

Corollary 1.13. Let H be a Hilbert space with an inner product (,) and $a(u,v)$ a bilinear continuous form on H satisfying for some $\nu > 0$

$$a(u,u) \geq \nu ||u||^2 \quad \forall u \in H. \tag{1.16}$$

Let K be a nonempty closed convex subset of H and $f \in H'$. Then there exists a unique u such that

$$u \in K, \quad a(u, v - u) \geq \langle f, v - u \rangle \quad \forall v \in K. \tag{1.6}'$$

Moreover, in the case $K = H$, u is the unique solution of

$$a(u,v) = \langle f, v \rangle \quad \forall v \in H. \tag{1.15}'$$

Proof: Denote by A the linear operator of $H \to H'$ such that

$$a(u,v) = \langle Au, v \rangle \quad \forall v \in H.$$

Then by (1.16), for all $v, v_0 \in K$ we have

$$\frac{\langle Av - Av_0, v - v_0 \rangle}{||v - v_0||} = \frac{\langle A(v-v_0), v-v_0 \rangle}{||v - v_0||} = \frac{a(v-v_0, v-v_0)}{||v - v_0||} \geq \nu ||v-v_0|| \to +\infty$$

and the result follows from Theorem 1.10.

(1.15)' follows immediately from Corollary 1.11.

Remark 1.14. The second part of this corollary is known as Lax-Milgram Theorem.

Comments

The first results about Variational inequalities were given by
Stampacchia [99] and Lions-Stampacchia [86], see also Fichera [58] and
Hartmann-Stampacchia [68]. The presentation that we have adopted here
borrows widely from Kinderlehrer-Stampacchia [76] (see also Brezis [27]
and the interesting introductions given in Baiocchi [14], Baiocchi-Capelo
[15] and Kinderlehrer [71]).

Some generalizations that will not be used here are possible. One
is, for instance, to find u such that

$$u \in K, \quad \langle Au, v - u \rangle + \phi(v) - \phi(u) \geq \langle f, v - u \rangle \quad \forall v \in K$$

where ϕ is a convex, weakly lower semi-continuous function (see Moreau
[90], Brezis [27], Lions [83]). Another one is to allow the set K to
depend on u; the problem is then called a quasi-variational inequality
(see Bensoussan-Lions [18], Lions [84], Tartar [105], Baiocchi-Capelo
[15]). See also Lions [83] and Brezis [27] for a generalization of
Corollary 1.11 and Mosco [91] for a different point of view of variational
inequalities.

Chapter 2
Examples and Applications

2.1. Some Functional Analysis

Let Ω be a bounded connected open subset of $\mathbb{R}^n$.

By $L^p(\Omega)$, we denote the usual spaces of "equivalence classes" of real functions whose p-th power is integrable $(1 \leq p < +\infty)$. $L^\infty(\Omega)$ will denote the space of functions which are essentially bounded. The norms on these spaces will be

$$|u|_p = \left(\int_\Omega |u(x)|^p dx \right)^{1/p} \qquad (1 \leq p < +\infty)$$

$$|u|_\infty = \text{ess sup } |u(x)|. \atop x \in \Omega$$

$$(2.1)$$

For k a positive integer, we shall denote by $W^{k,p}(\Omega)$, the Sobolev space defined for every $1 \leq p \leq +\infty$ by

$$W^{k,p}(\Omega) = \{u \in L^p(\Omega) \,|\, D^\alpha u \in L^p(\Omega) \quad \forall \, |\alpha| \leq k\} \qquad (2.2)$$

where in (2.2) $D^\alpha u$ denotes the derivative $\dfrac{\partial^{|\alpha|}}{\partial x_1^{\alpha_1} \ldots \partial x_n^{\alpha_1}}$ in the distributional sense $(|\alpha| = \alpha_1 + \ldots + \alpha_n)$.

It is well known that these spaces are Banach spaces with the norm

$$||u||_{k,p} = \sum_{|\alpha| \leq k} |D^\alpha u|_p. \qquad (2.3)$$

In the case $k = 1$, we will also use the spaces $W_0^{1,p}(\Omega)$, defined for $1 \leq p < +\infty$ by

$$W_0^{1,p}(\Omega) = \text{the closure of } \mathscr{D}(\Omega) \text{ in } W^{1,p}(\Omega), \qquad (2.4)$$

where $\mathscr{D}(\Omega)$ is the space of infinitely differentiable functions with compact support in Ω.

On these spaces, we often use in place of (2.3), the norm

$$||\nabla u||_p \tag{2.5}$$

where $\nabla = (\frac{\partial}{\partial x_1}, \ldots, \frac{\partial}{\partial x_n})$ and $|\ |$ is the usual Euclidean norm in $\mathbb{R}^n$, i.e., we have:

$$|\nabla u| = \left(\sum_{i=1}^{n} (\frac{\partial u}{\partial x_i})^2 \right)^{1/2}. \tag{2.6}$$

The fact that (2.3) and (2.5) are equivalent is an easy consequence of the following theorem:

Theorem 2.1. (Poincaré's Inequality). There is a constant C depending only on Ω such that

$$|u|_p \leq C ||\nabla u||_p \qquad \forall u \in W_0^{1,p}(\Omega). \tag{2.7}$$

Proof: By density, it is enough to prove (2.7) for u in $\mathscr{D}(\Omega)$. Since Ω is bounded, we have $\Omega \subset [-a,a] \times \mathbb{R}^{n-1}$ for some $a > 0$. Assume that we have extended u to be 0 outside Ω, then we have

$$u(x_1, \ldots, x_n) = \int_{-a}^{x_1} \frac{\partial u}{\partial x_1}(t, x_2, \ldots, x_n) dt$$

and by Hölder's inequality, we get in Ω

$$|u(x_1, \ldots, x_n)| \leq \int_{-a}^{+a} |\frac{\partial u}{\partial x_1}(t, x_2, \ldots, x_n)|^p dt^{1/p} (2a)^{1/q}$$

with $\frac{1}{p} + \frac{1}{q} = 1$. But now, taking the power p of this inequality and integrating over Ω leads to

$$|u|_p^p \leq (2a)^{\frac{p}{q}+1} |\frac{\partial u}{\partial x_1}|_p^p$$

and so

$$|u|_p \leq 2a |\frac{\partial u}{\partial x_1}|_p \leq 2a ||\nabla u||_p \qquad \forall u \in \mathscr{D}(\Omega)$$

which is (2.7). (Note that we have only used the fact that Ω is bounded in one direction.)

As an improvement of (2.7), we have in fact (see [65] or [95] for a proof).

Theorem 2.2. (Sobolev).

$$W_0^{1,p}(\Omega) \subset \begin{cases} L^{p^*}(\Omega) & \text{for } p < n \\ \\ C_0(\overline{\Omega}) & \text{for } p > n \end{cases}$$

where p^* is defined by $\dfrac{1}{p^*} = \dfrac{1}{p} - \dfrac{1}{n}$.

Moreover, there exists a constant C which depends only on n, p, Ω such that

$$|u|_{p^*} \leq C||\nabla u||_p \quad \text{for } p < n \tag{2.8}$$

$$|u|_\infty \leq C||\nabla u||_p \quad \text{for } p > n. \tag{2.9}$$

$(C_0(\overline{\Omega})$ denotes the space of continuous function on Ω which vanish on the boundary. We will sometimes use more general embedding theorems. We refer to [1] or [95] for this matter).

We will always use the notations $H^1(\Omega)$ and $H_0^1(\Omega)$ for $W^{1,2}(\Omega)$, $W_0^{1,2}(\Omega)$, respectively, and $H^{-1}(\Omega)$ for the dual of $H_0^1(\Omega)$.

For $f_0, f_1, \ldots, f_n \in L^2(\Omega)$, the mapping

$$T : v \longmapsto \langle T, v \rangle = \int_\Omega f_0 v - f_i \frac{\partial v}{\partial x_i} \tag{2.10}$$

(As it will be done often in the sequel, we have ommitted here the measure dx and we have used the summation convention - i.e., $f_i \dfrac{\partial v}{\partial x_i}$ means $\sum_i f_i \dfrac{\partial v}{\partial x_i}$) gives us an element of $H^{-1}(\Omega)$ that we write as:

$$T = f_0 + \frac{\partial f_i}{\partial x_i}. \tag{2.11}$$

Conversely, if $T \in H^{-1}(\Omega)$ by the Riez representation theorem, we can find $u \in H_0^1(\Omega)$ such that

$$\langle T, v \rangle = \int_\Omega u \cdot v + \nabla u \cdot \nabla v$$

and T is of type (2.11). More generally, if we denote by $W^{-1,q}(\Omega)$ the dual of $W_0^{1,p}(\Omega)$ $(1 < p < +\infty, \frac{1}{p} + \frac{1}{q} = 1)$, we can prove that all elements of $W^{-1,q}$ are given by (2.10), (2.11) with $f_i \in L^q(\Omega)$. Moreover, $W_0^{1,p}(\Omega)$ is a reflexive Banach space (see [82]). For $T_1, T_2 \in W^{-1,q}(\Omega)$, we sometimes will use the notation

$$T_1 \geq T_2 \tag{2.12}$$

which simply means that

$$\langle T_1 - T_2, v \rangle \geq 0 \qquad \forall v \in W_0^{1,p}(\Omega), \quad v \geq 0 \quad \text{a.e.} \tag{2.13}$$

In other words, $T_1 - T_2$ is a positive measure.

At this point, these facts are sufficient for our purposes in this section. However, let us quote some other useful results which will be used later on. First:

Theorem 2.3. Let $u \in W^{1,p}(\Omega)$ and f be a piecewise smooth function on $\mathbb{R}$ with $f' \in L^\infty(\mathbb{R})$. Then if L denotes the set of corner points of f, we have in the distributional sense

$$\nabla(f \circ u) = \begin{cases} f'(u) \cdot \nabla u & \text{if } u \notin L \\ 0 & \text{if } u \in L \end{cases}. \tag{2.14}$$

Proof: See [65] or [100].

If we denote by $u^+ = \text{Max}(u,0)$ (Resp. $-u^- = \text{Min}(0,u)$), the positive (Resp. negative) part of a function u, we deduce easily from Theorem 2.3 that:

Theorem 2.4. If $u \in W^{1,p}(\Omega)$, then u^+, u^- are in $W^{1,p}(\Omega)$. Moreover, we have

$$\nabla u^+ = \begin{cases} \nabla u & \text{on } u > 0 \\ 0 & \text{on } u \leq 0 \end{cases} \tag{2.15}$$

$$\nabla u^- = \begin{cases} 0 & \text{on } u \geq 0 \\ -\nabla u & \text{on } u < 0 \end{cases} \tag{2.16}$$

in the distributional sense.

If $u \in W_0^{1,p}(\Omega)$, then so are u^+ and u^- and more generally $(u-k)^+$, $(u+k)^-$ for any positive constant k.

Proof: Apply the preceding theorem with $f(t) = \max(0,t)$ and $\min(0,t)$ to get (2.15), (2.16) and approximate u by a smooth function to get the second part.

Remark 2.5. As a consequence, we see that for a function u in $W^{1,p}(\Omega)$, $\nabla u = 0$ a.e. on the set $[u = 0]$.

Finally, we quote:

Theorem 2.5. $H_0^1(\Omega)$ is compactly embedded in $L^2(\Omega)$ (i.e., the identity is compact from $H_0^1(\Omega)$ in $L^2(\Omega)$) and if the boundary Γ of Ω is Lipschitz, $H^1(\Omega)$ is compactly embedded in $L^2(\Omega)$.

Proof: See [65] and [1] where much more is proved.

In the sequel, we will use the following notations:

$C(\Omega)$, $C^k(\Omega)$ for the spaces of continuous, k times derivable functions on Ω,

$C(\overline{\Omega})$ for the space of continuous functions on $\overline{\Omega}$ the closure of Ω,

$C^k(\overline{\Omega})$ for the space of functions in $C^k(\Omega)$ whose derivatives can be extended up to order k to continuous functions on $\overline{\Omega}$

$C^{0,\alpha}(\overline{\Omega})$ for the space of Hölder continuous functions in $\overline{\Omega}$ of order α,

$C^{k,\alpha}(\overline{\Omega})$ for the space of functions in $C^k(\overline{\Omega})$ whose derivatives of order k are in $C^{0,\alpha}(\overline{\Omega})$.

2.2. The Dirichlet Problem

As a first simple result, we have (if Δ denotes the usual Laplacian):

<u>Theorem 2.6.</u> Let $f \in H^{-1}(\Omega)$, then there exists a unique solution u of the Dirichlet problem:

$$\begin{cases} -\Delta u = f \\ u \in H_0^1(\Omega). \end{cases} \tag{2.17}$$

<u>Proof:</u> Apply Corollary 1.13 with $X = H_0^1(\Omega)$, $a(u,v) = \int_\Omega \nabla u \cdot \nabla v \, dx$. Note that in this case, (2.17) is just the Riez representation theorem.

More generally, let a_{ij} (i,j = 1,...,n) be functions in $L^\infty(\Omega)$ such that (with the summation convention)

$$a_{ij}(x)\xi_i\xi_j \geq \nu|\xi|^2 \qquad \forall x \in \Omega, \quad \forall \xi \in \mathbb{R}^n \tag{2.18}$$

where ν is some strictly positive constant and $|\xi|$ the Euclidean norm of ξ defined by

$$|\xi|^2 = \sum_{i=1}^n \xi_i^2, \qquad \forall \xi \in \mathbb{R}^n. \tag{2.19}$$

Then

$$A = \frac{\partial}{\partial x_i}\left(a_{ij}(x)\frac{\partial}{\partial x_j}\right) \tag{2.20}$$

defines an operator from $H_0^1(\Omega)$ in $H^{-1}(\Omega)$. Indeed, (see (2.10), (2.11)), for $u \in H_0^1(\Omega)$, we define $-Au$ by the formula

$$<-Au \cdot v> = \int_\Omega a_{ij}(x)\frac{\partial u}{\partial x_i} \cdot \frac{\partial v}{\partial x_j} \, dx \qquad \forall v \in H_0^1(\Omega). \tag{2.21}$$

(In the sequel, the integral of the right side of this equality will always be written as $<-Au,v>$.)

Assume now that $F(x,y)$ is a function on $\Omega \times \mathbb{R}$ satisfying

F is measurable in x for all y and continuous in y for
almost all x in Ω .   (2.22)

There exists two positive functions $C(x) \in L^{\infty}(\Omega)$,
$C'(x) \in L^2(\Omega)$ such that

$$|F(x,y)| \leq C(x)|y| + C'(x).$$ (2.23)

F is monotone in y , i.e.,

$$[F(x,y)-F(x,y')][y-y'] \geq 0 \quad \forall y,y' \in \mathbb{R}, \quad \text{a.e. in } \Omega.$$ (2.24)

Then we have:

Theorem 2.7. Under the above assumptions, for $f \in H^{-1}(\Omega)$ there exists
a unique solution u of the nonlinear Dirichlet problem:

$$\begin{cases} -Au + F(x,u) = f \quad \text{in } \mathscr{D}'(\Omega) \quad (\text{or } H^{-1}(\Omega)) \\ u \in H_0^1(\Omega). \end{cases}$$ (2.25)

Proof: Note that by (2.22), (2.23), we have for all $u \in L^2(\Omega)$, $F(x,u) \in L^2(\Omega) \subset H^{-1}(\Omega)$. Now it suffices to apply the Corollary 1.11 with the
operator

$$u \longmapsto -Au + F(x,u).$$

Indeed from (2.18), (2.24), we get

$$<[-Au + F(x,u)] - [-Av + F(x,v)], u - v>$$
$$\geq <-A(u - v), u - v> \geq \nu ||\nabla(u - v)||_2^2$$

which concludes the proof.

Remark 2.8. In particular, the above theorem gives us existence and uni-
ueness for the solution of nonlinear Dirichlet problems of the type

$$u \in H_0^1(\Omega), \quad -Au + F(u) = f$$

provided F is monotone and bounded by $|F(u)| \leq C|u| + C'$. For further
extensions see [26].

2.3. The Obstacle Problems

Let $a_{ij}(x)$ be $L^\infty(\Omega)$-functions satisfying (2.18) and A defined by (2.20), (2.21). For ϕ,ψ measurable functions in Ω set

$$K_\phi = \{v \in H^1_0(\Omega) | v(x) \geq \phi(x) \quad \text{a.e. in} \Omega\} \tag{2.26}$$

$$K_\phi^\psi = \{v \in H^1_0(\Omega) | \psi(x) \geq v(x) \geq \phi(x) \quad \text{a.e. in} \Omega\}. \tag{2.27}$$

Then we have

Theorem 2.9. If $K_\phi \neq \emptyset$ (Resp. $K_\phi^\psi \neq \emptyset$), then for $f \in H^{-1}(\Omega)$, there exists a unique solution u_1 (Resp. u_2) of the problem

$$u_1 \in K_\phi, \ <-Au_1, v - u_1> \geq <f, v - u_1> \quad \forall v \in K_\phi \tag{2.28}$$

(Resp.

$$u_2 \in K_\phi^\psi, \ <-Au_2, v - u_2> \geq <f, v - u_2> \quad \forall v \in K_\phi^\psi). \tag{2.29}$$

u_1 and u_2 are called the solutions of a one obstacle problem and of a two obstacle problem respectively.

Proof: It is easy to check that K_ϕ and K_ϕ^ψ are closed convex sets in $H^1_0(\Omega)$, so the result follows immediately from Corollary 1.13, with

$$a(u,v) = <-Au, v> = \int_\Omega a_{ij} \frac{\partial u}{\partial x_i} \frac{\partial v}{\partial x_j} \, dx$$

(see (2.21)).

Remark 2.10. The physical interpretation of (2.28) (for example) is the following. Assume that $\Omega \subset \mathbb{R}^2$ represents a membrane attached to $\Gamma = \partial\Omega$. The shape of the solid obstacle which pushes on the membrane is described by ϕ. Then the equilibrium position of the membrane is given by the

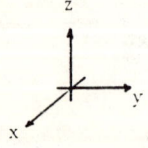
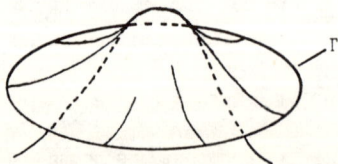

solution of

$$u_1 \in K_\phi, \int_\Omega \nabla u_1 \cdot \nabla(v - u_1) dx \geq 0 \quad \forall v \in K_\phi.$$

Remark 2.11.   We will discuss obstacle problems involving nonlinear operators in Section 2.5.   We could also consider convex K_ϕ with nonvanishing
boundary values (see, for instance, Chapter 4).

2.4.   Elastic Plastic Torsion Problems

In this section, we present some problems involving constraints on
the gradient.   More precisely, set

$$K = \{v \in H_0^1(\Omega) \mid |\nabla v(x)| \leq 1 \quad \text{a.e. on} \quad \Omega\}. \tag{2.30}$$

K is a closed convex set of $H_0^1(\Omega)$ which contains 0, and so, under the
assumption of the previous sections on the a_{ij}, we get as an application
of Corollary 1.13:

Theorem 2.12.   For $f \in H^{-1}(\Omega)$, there exists a unique solution u of

$$u \in K, \quad <-Au, v - u> \geq <f, v - u> \quad \forall v \in K. \tag{2.31}$$

Remark 2.13.   When $a_{ij} = \delta_{ij}$ and $f = \mu = \text{const.}$ the problem is reduced
to

$$u \in K, \quad \int_\Omega \nabla u \cdot \nabla (v - u) dx \geq \int_\Omega \mu \cdot (v - u) dx \quad \forall v \in K. \tag{2.32}$$

This is known as the elastic-plastic torsion problem.   It arises when
studying the torsion of a cylindrical bar of section $\Omega \subset \mathbb{R}^2$.   The bar is
made of an elastic-plastic material (see [77]), μ is a constant propor-
tional to the angle of twist of the end section of the bar which is not
clamped.

This problem has given rise to many investigations during the past
decade, see [77], [78], [42], [43], [63], and [106].

More generally, assume that $\mathscr{C}_1, \mathscr{C}_2, \ldots, \mathscr{C}_p$ are disjoint open sets
in Ω satisfying

$$\mathscr{C}_i \subset \Omega \quad \forall i, \quad \overline{\mathscr{C}}_i \cap \overline{\mathscr{C}}_j = \emptyset \quad \forall i \neq j.$$

(The different $\mathscr{C}_i$ are for instance the sections of cylindrical cavities
in a bar in torsion - see the figure on the following page.)
Then instead of K one can consider

$$K' = \{v \in H_0^1(\Omega) \mid |\nabla(x)| \leq 1 \quad \text{a.e. in} \quad \Omega,$$
$$v = C_i = \text{const. on} \quad \mathscr{C}_i\} \tag{2.33}$$

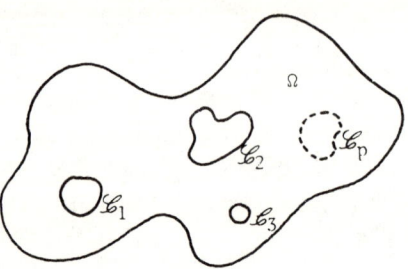

(the constants C_i are not prescribed, but depend on the function v).
Then under the same assumptions as above, we have that there exists a unique
solution u of (2.31) with K' in place of K. One has just to prove
that K' is a closed convex set of $H_0^1(\Omega)$. (See [78], [42], and [107].)

2.5. Nonlinear Operators

Let $a_1(x),\ldots,a_n(x)$ be functions in $L^\infty(\Omega)$ satisfying

$$a_i(x) \geq \nu \quad \forall i = 1,\ldots,n \tag{2.34}$$

for some $\nu > 0$.

Given two measurable functions ϕ,ψ, let us define

$$K_\phi = \{v \in W_0^{1,p}(\Omega) \mid v(x) \geq \phi(x) \quad \text{a.e. on } \Omega\}$$
$$K_\phi^\psi = \{v \in W_0^{1,p}(\Omega) \mid \psi(x) \geq v(x) \geq \phi(x) \quad \text{a.e. on } \Omega\}$$

with $p \geq 2$.

Then we have:

Theorem 2.14. Assume that K_ϕ (Resp. K_ϕ^ψ) is nonempty and that (2.34)
holds. Then for $f \in W^{-1,q}(\Omega)$ $(\frac{1}{q} + \frac{1}{p} = 1)$, there exists a unique solution u_1 (Resp. u_2) of

$$u_1 \in K_\phi, \quad \int_\Omega a_i(x) \left|\frac{\partial u_1}{\partial x_i}\right|^{p-2} \cdot \frac{\partial u_1}{\partial x_i} \cdot \frac{\partial}{\partial x_i}(v - u_1)\,dx$$

$$\geq \langle f, v - u_1 \rangle \quad \forall v \in K_\phi. \tag{2.35}$$

(Resp.:

$$u_2 \in K_\phi^\psi, \quad \int_\Omega a_i(x) \left|\frac{\partial u_2}{\partial x_i}\right|^{p-2} \cdot \frac{\partial u_2}{\partial x_i} \cdot \frac{\partial}{\partial x_i}(v - u_2)\,dx$$

$$\geq \langle f, v - u_2 \rangle \quad \forall v \in K_\phi^\psi.) \tag{2.36}$$

<u>Proof</u>: Let A be the operator defined from $W_0^{1,p}(\Omega)$ into $W^{-1,q}(\Omega)$ by:

$$Au = \frac{\partial}{\partial x_i}\left(a_i(x)\left|\frac{\partial u}{\partial x_i}\right|^{p-2}\frac{\partial u}{\partial x_i}\right).$$

(It is easy to see that $a_i(x)\left|\frac{\partial u}{\partial x_i}\right|^{p-2}\cdot\frac{\partial u}{\partial x_i} \in L^q(\Omega)$ when $\frac{\partial u}{\partial x_i} \in L^p(\Omega)$, $\frac{1}{p} + \frac{1}{q} = 1$.)

For this operator, we have

$$<-(Au - Av),u - v> = \int_\Omega a_i(x)\left(\left|\frac{\partial u}{\partial x_i}\right|^{p-2}\cdot\frac{\partial u}{\partial x_i} - \left|\frac{\partial v}{\partial x_i}\right|^{p-2}\cdot\frac{\partial v}{\partial x_i}\right)\cdot$$

$$\cdot\left(\frac{\partial u}{\partial x_i} - \frac{\partial v}{\partial x_i}\right) \geq 0$$

by (2.34) and by monotonicity of the mapping $x \to |x|^{p-2}x$. Thus $-A$ is monotone and even strictly monotone (by (2.34) and the strict monotonicity of $x \to |x|^{p-2}x$).

Now by Hölder's inequality, we get (note that $a_i \in L^\infty(\Omega)$)

$$|<Au,v>| = \left|\int_\Omega a_i(x)\left|\frac{\partial u}{\partial x_i}\right|^{p-2}\cdot\frac{\partial u}{\partial x_i}\cdot\frac{\partial v}{\partial x_i}\right|$$

$$\leq C\left(\int_\Omega\left|\frac{\partial u}{\partial x_i}\right|^{p-1\cdot q}\right)^{1/q}\cdot\left(\int\left|\frac{\partial v}{\partial x_i}\right|^p\right)^{1/p}.$$

Hence

$$|<-Au,v>| \leq C||\nabla u||_p^{p/q}||\nabla v||_p \qquad \forall u,v \in W_0^{1,p}(\Omega) \tag{2.37}$$

and so if v_0 is an element in K (or in $W_0^{1,p}(\Omega)$), we get from this inequality

$$<-(Au-Av_0),u - v_0> = <-Au,u> - <-Av_0,u - v_0> - <-Au,v_0>$$

$$\geq \nu||\nabla u||_p^p - C||\nabla(u - v_0)||_p - C||\nabla u||_p^{p/q}$$

$$\geq \nu||\nabla u||_p^p - C||\nabla u||_p - C||\nabla u||_p^{p/q}$$

$$- C||\nabla v_0||_p \qquad \forall u \in W_0^{1,p}(\Omega).$$

Since $\frac{1}{p} + \frac{1}{q} = 1$, we have $\frac{p}{q} = p - 1 < p$ and thus

$$<-(Au - Av_0),u - v_0>/||\nabla u||_p \to +\infty \quad \text{as} \quad ||\nabla u||_p \to +\infty.$$

This verifies the coerciveness assumption (1.9) for $-A$ (note that $||\nabla(u-v_0)||_p \sim ||\nabla u||_p$ when $||\nabla u||_p \to +\infty$), and the result follows from

Theorem 1.10, since K_ϕ and K_ϕ^ψ are closed and convex.

As a second example, one can consider for $1 < \alpha \leq 2$

$$a_i(\xi) = (1 + |\xi|^2)^{\alpha-2/2} \, \xi_i \qquad \forall \xi \in \mathbb{R}^n. \tag{2.38}$$

We define an operation from $W_0^{1,\alpha}(\Omega)$ into $W^{-1,\alpha'}(\Omega)$ $(\frac{1}{\alpha} + \frac{1}{\alpha'} = 1)$ by setting

$$Au = \frac{\partial}{\partial x_i} (a_i(\nabla u)).$$

Then given $f \in W^{-1,\alpha'}(\Omega)$, there exists a unique solution u_1 of

$$u_1 \in K_\phi, \quad \int_\Omega a_i(u_1)\frac{\partial}{\partial x_i}(v - u_1)dx \geq <f,v - u_1> \quad \forall v \in K_\phi \tag{2.39}$$

with $K_\phi = \{v \in W_0^{1,\alpha}(\Omega) \,|\, v(x) \geq \phi(x)$ a.e. in $\Omega\}$ provided that the closed convex set K_ϕ is not empty (see [76] for a proof).

Let us note that when $\alpha = 1$, A is the prescribed mean curvature operator, and it is known that even for the Dirichlet problem, the existence of a solution may fail. When Ω is smooth and convex, $\phi \in W^{1,\infty}(\Omega)$, $\phi < 0$ on Γ, the interested reader will find a proof of existence and uniqueness of the solution of (2.39) in [76].

2.6. Fourth Order Variational Inequalities

Recently (see [38], [44]) variational inequalities with fourth order operators have been investigated. For instance, if $f \in H^{-1}(\Omega)$ one can prove that there exists a unique solution u of (see [38])

$$u \in K, \quad \int_\Omega \Delta u \cdot \Delta(v - u)dx \geq <f,v - u> \quad \forall v \in K$$

with, for example, K equal to

$$K = \{u \in H_0^1(\Omega) \cap W^{2,2}(\Omega) \,|\, |\Delta u(x)| \leq \alpha \text{ a.e. in } \Omega\},$$

where α is some positive constant.

More generally, if Q is a positive quadratic form of n^2 variables, one can consider, for $\alpha > 0$,

$$K = \{u \in H_0^1(\Omega) \cap W^{2,2}(\Omega) \,|\, Q(u_{x_i x_j}) \leq \alpha\}.$$

A relevant physical example is given in $\mathbb{R}^2$ (see [79]) by

$$K = \{u \in H_0^1(\Omega) \cap W^{2,2}(\Omega) \,|\, (u_{xx} - u_{yy})^2 + 4u_{xy}^2 \leq 1\}.$$

Comments

The reader interested in more results on the Sobolev spaces is referred to Adams [1]. See also Necas [95].

We have restricted our examples to elliptic problems and elliptic
variational inequalities in divergence form: For parabolic variational
inequalities see Mignot-Puel [89] or Lions [84], Bensoussan-Lions [19];
for problems which are not on divergence form see Bensoussan-Lions [19]
or P. L. Lions [87]. The relationship between variational inequalities
and convex programming are investigated in Mancino-Stampacchia [88]. For
nonlinear variational inequalities, see, for instance, M. F. Bidaut-Veron
[21] and the bibliography of this paper.

In this chapter we have restricted our examples to the simplest and,
maybe, most famous ones. Many other applications related to mechanics or
optimal control theory can be found in [19], [57], [76] and in the two
volumes of [13].

Note also that throughout these notes we will not discuss the numerical analysis of the problems investigated. Much of this material is covered by the books:

R. Glowinski, J. L. Lions, R. Tremolières: Numerical Analysis of
Variational Inequalities, North Holland, 1981.

P. G. Ciarlet: The Finite Element Method for Elliptic Problems,
North Holland, 1978.

Chapter 3

The Obstacle Problems:
A Regularity Theory

Throughout this chapter, Ω will be a bounded domain of $\mathbb{R}^n$ with boundary Γ and a_{ij} functions in $L^\infty(\Omega)$ which satisfy the ellipticity assumption

$$a_{ij}(x)\xi_i\xi_j \geq \nu|\xi|^2 \qquad \forall x \in \Omega, \quad \forall \xi \in \mathbb{R}^n. \tag{3.1}$$

We will denote by A the operator

$$A = \frac{\partial}{\partial x_i}\left(a_{ij}(x)\frac{\partial}{\partial x_j}\right). \tag{3.2}$$

3.1. Monotonicity Results

For $f, f' \in H^{-1}(\Omega), \phi, \phi'$ measurable functions, let us denote by u_1 and u_1' the solutions of

$$u_1 \in K_\phi, \quad <-Au_1, v - u_1> \geq <f, v - u_1> \quad \forall v \in K_\phi \tag{3.3}$$

$$u_1' \in K_{\phi'}, \quad <-Au_1', v - u_1'> \geq <f', v - u_1'> \quad \forall v \in K_{\phi'}, \tag{3.4}$$

where

$$K_\phi = \{v \in H_0^1(\Omega) \,|\, v(x) \geq \phi(x) \quad \text{a.e. in } \Omega\}$$

and

$$K_{\phi'} = \{v \in H_0^1(\Omega) \,|\, v(x) \geq \phi'(x) \quad \text{a.e. in } \Omega\}$$

are not empty.

Then we have

Proposition 3.1: Assume that $f \geq f'$ (in the H^{-1} sense) and $\phi(x) \geq \phi'(x)$ a.e., then $u_1(x) \geq u_1'(x)$ a.e.

Proof: (See Brezis [28]). We have $(u_1 - u_1')^- \in H_0^1(\Omega)$ (see Theorem 2.4) and it is easy to check that

$$u_1 + (u_1 - u_1')^- \in K_\phi, \quad u_1' - (u_1 - u_1')^- \in K_{\phi'}.$$

Using (3.3) and (3.4), this leads to

$$<-Au_1,(u_1 - u_1')^-> \geq <f,(u_1 - u_1')^->$$

$$<-Au_1',-(u_1 - u_1')^-> \geq <f',-(u_1 - u_1')>.$$

Now by adding the above inequalities, we obtain

$$<-A(u_1 - u_1'),(u_1 - u_1')^-> \geq <f - f',(u_1 - u_1')^-> \geq 0.$$

That is to say, (see (3.1), (2.16))

$$\nu||\nabla(u_1 - u_1')^-||_2^2 \leq -<-A(u_1 - u_1'),(u_1 - u_1')^-> \leq 0$$

and so $(u_1 - u_1')^- = 0$, which gives the result.

With a similar proof, one can show:

Proposition 3.2. Assume that $\phi,\phi' \in L^\infty(\Omega)$. If K_ϕ and $K_{\phi'}$ are non-empty and if u_1,u_1' denote respectively the solutions of

$$u_1 \in K_\phi, \quad <Au_1,v - u_1> \geq 0 \quad \forall v \in K_\phi \tag{3.5}$$

$$u_1' \in K_{\phi'}, \quad <-Au_1',v - u_1'> \geq 0 \quad \forall v \in K_{\phi'}, \tag{3.6}$$

then we have

$$0 \leq u_1 \leq |\phi^+|_\infty \tag{3.7}$$

$$|u_1 - u_1'|_\infty \leq |\phi - \phi'|_\infty. \tag{3.8}$$

Proof: For the first inequality, let $k = |\phi^+|_\infty$ and take in (3.5)

$$v = u_1^+ \geq u_1 \geq \phi \quad \text{and} \quad v = u_1 - (u_1 - k)^+ = \min(u_1,k) \geq \phi,$$

which leads to

$$0 \leq <-Au_1,u_1^+ - u_1> = -<-Au_1^-,u_1^-> \Rightarrow u_1^- = 0 \Longleftrightarrow u_1 \geq 0$$

$$0 \leq <-Au_1,-(u_1,-k)^+> = -<A(u_1 - k)^+,(u_1 - k)^+> \Rightarrow (u_1 - k)^+ = 0 \Longleftrightarrow u_1 \leq k.$$

The other inequality follows in the same way, and the proof is left to the reader.

Remark 3.3.   One can, of course, give similar results for the two obstacles problems. Reasonable assumptions are found by considering the physical problem as stated in Remark 2.10. Indeed, Proposition 3.1 (with f = f' = 0) means only that the more the membrane is pushed, the more it goes up.

3.2.   The Penalty Method

Let β_1, β_2 be two functions satisfying

$$\beta_1 \text{ is Lipschitz, monotone, } \beta_1(t) = 0 \qquad \forall t \geq 0 \qquad (3.9)$$

$$\beta_2 \text{ is Lipschitz, monotone, } \beta_2(t) = 0 \qquad \forall t \leq 0. \qquad (3.10)$$

Then for $\phi, \psi \in L^2(\Omega)$, $f \in H^{-1}(\Omega)$ and by Theorem (2.7), for all $\epsilon > 0$, there exists a unique solution u_ϵ of:

$$\begin{cases} -Au_\epsilon + \dfrac{\beta_1}{\epsilon}(u_\epsilon - \phi) + \dfrac{\beta_2}{\epsilon}(u_\epsilon - \psi) = f \\ u_\epsilon \in H_0^1(\Omega). \end{cases} \qquad (3.11)$$

(Indeed, from the Lipschitz character of β_i and since $\beta_i(0) = 0$, we have, for some constants C_1, C_2 ,

$$|F(x,y)| = |\beta_1(y-\phi) + \beta_2(y-\psi)| \leq C_1|y - \phi(x)| + C_2|y - \psi(x)|$$

$$\leq C|y| + C'(|\phi(x)| + |\psi(x)|).)$$

Now we want to show that for a suitable choice of β_1, β_2 and ϵ small enough, u_ϵ provides us with an approximation of u_1 or u_2 , the solutions of

$$u_1 \in K_\phi, \quad <-Au_1, v - u_1> \geq <f, v - u_1> \quad \forall v \in K_\phi \qquad (3.12)$$

$$u_2 \in K_\phi^\psi, \quad <-Au_2, v - u_2> \geq <f, v - u_2> \quad \forall v \in K_\phi^\psi. \qquad (3.13)$$

As we will see later, this approximation is very convenient in regularity theory, since it allows us to deal with an equality instead of an inequality.

So we have

Theorem 3.4.   (i) If $K_\phi \neq \emptyset$, $\beta_1(t) < 0 \ \forall t < 0$, $\beta_2 \equiv 0$, then when $\epsilon \to 0$, $u_\epsilon \to u_1$ in $H_0^1(\Omega)$, u_1 being the solution of (3.12).

(ii) If $K_\phi^\psi \neq \emptyset$, $\beta_1(t) < 0 \ \forall t < 0$, $\beta_2(t) > 0 \ \forall t > 0$, then when $\epsilon \to 0$, $u_\epsilon \to u_2$ in $H_0^1(\Omega)$, u_2 being the solution of (3.13).

Proof: Choose v in K_ϕ in case (i) and v in K_ϕ^ψ in case (ii). For such a v, we have by (3.9), (3.10)

$$\beta_1(v - \phi) + \beta_2(v - \psi) = 0$$

and by the monotonicity assumptions on the β_i , we get

$$\int_\Omega (\beta_1(u_\varepsilon-\phi) + \beta_2(u_\varepsilon-\psi))\cdot(u_\varepsilon-v) \geq \int_\Omega (\beta_1(v-\phi) + \beta_2(v-\psi))\cdot(u_\varepsilon-v) = 0.$$

But by (3.11), this can also be written as

$$<-Au_\varepsilon, u_\varepsilon - v> \leq <f, u_\varepsilon - v> \tag{3.14}$$

and after an easy computation, using (3.2) and Theorem 2.1, we get

$$\nu||u_\varepsilon||_{1,2}^2 \leq C||u_\varepsilon||_{1,2} + C'$$

where C and C' are constants which don't depend on ε . So, we have
for some constant C

$$||u_\varepsilon||_{1,2} \leq C, \tag{3.15}$$

by Theorem 2.5, we can construct a sequence $\varepsilon_n \to 0$ such that

$$u_{\varepsilon_n} \longrightarrow u \text{ in } H_0^1(\Omega) \text{ weakly}$$

$$u_{\varepsilon_n} \longrightarrow u \text{ in } L^2(\Omega) \text{ strongly.}$$

By Lebesgue's Theorem and from the Lipschitz character of the β_i , we
deduce from $u_{\varepsilon_n} \longrightarrow u$ in $L^2(\Omega)$ that

$$\beta_1(u_{\varepsilon_n}-\phi) + \beta_2(u_{\varepsilon_n}-\psi) \to \beta_1(u-\phi) + \beta_2(u-\psi) \text{ in } L^2(\Omega) \text{ (and so in } \mathscr{D}'(\Omega)).$$

But by (3.11), (3.15), we also have

$$\beta_1(u_{\varepsilon_n}-\phi) + \beta_2(u_{\varepsilon_n}-\psi) = \varepsilon_n(Au_{\varepsilon_n}+f) \to 0 \text{ in } \mathscr{D}'(\Omega)$$

and by the uniqueness of the limit in $\mathscr{D}'(\Omega)$, we get

$$\beta_1(u-\phi) + \beta_2(u-\psi) = 0,$$

which gives $u \in K_\phi$ in case (i) and $u \in K_\phi^\psi$ in case (ii). Since (3.14)
is true for all v in K_ϕ in case (i) and all v in K_ϕ^ψ in case (ii),
we get

$$<-Au_{\varepsilon_n}, v - u_{\varepsilon_n}> \geq <f, v - u_{\varepsilon_n}> \quad \forall v \in K_\phi \text{ (Resp. } v \in K_\phi^\psi).$$

By the monotonicity of $-A$

$$<-Av, v - u_{\epsilon_n}> \; \geq \; <f, v - u_{\epsilon_n}> \quad \forall v \in K_\phi \quad (\text{Resp. } v \in K_\phi^\psi).$$

Letting ϵ_n go to 0 leads to

$$<-Av, v - u> \; \geq \; <f, v - u> \quad \forall v \in K_\phi \quad (\text{Resp. } v \in K_\phi^\psi).$$

By Minty's Lemma and since $u \in K_\phi$, we obtain that $u = u_1$ in case (i)
(Resp. $u = u_2$ in case (ii)). Now by uniqueness of u_1 and u_2 , we have
$u_\epsilon \longrightarrow u_i$ in $H_0^1(\Omega)$. To get strong convergence, it is enough to note
that (see (3.1), (3.14))

$$\nu||u - u_\epsilon||_{1,2}^2 \leq <-A(u - u_\epsilon), u - u_\epsilon> \; = \; <-Au, u - u_\epsilon> + <Au_\epsilon, u - u_\epsilon>$$

$$\leq \; <-Au, u - u_\epsilon> - <f, u - u_\epsilon> \to 0$$

with ϵ . This concludes the proof.

Remark 3.5. By choosing $\beta_1 \equiv 0$, $\beta_2(t) > 0$ $\forall t > 0$, we would obtain the
convergence of u_ϵ to the solution u^1 of:

$$u^1 \in K^\psi, \; <-Au^1, v - u^1> \; \geq \; <f, v - u^1> \quad \forall v \in K^\psi$$

under the assumption that $K^\psi = \{v \in H_0^1(\Omega) | v(x) \leq \psi(x) \quad \text{a.e.}\} \neq \emptyset$.

Remark 3.6. It is easy to see that the above proof also holds for a nonlinear operator A from $H_0^1(\Omega)$ into $H^{-1}(\Omega)$ which satisfies the assumptions of Minty's Lemma and the coercivity property

$$<Au - Av, u - v> \; \geq \; \nu||u - v||_{1,2} \quad \forall u, v \in H_0^1(\Omega).$$

Remark 3.7. In the next part, we will use u_ϵ to denote the solution of
a slightly more complicated problem, that is to say

$$\begin{cases} -Au_\epsilon + \dfrac{\beta_1}{\epsilon}(u_\epsilon - \phi) + \epsilon(u_\epsilon - \phi) + \dfrac{\beta_2}{\epsilon}(u_\epsilon - \psi) + \epsilon(u_\epsilon - \psi) = f \\ u_\epsilon \in H_0^1(\Omega). \end{cases} \qquad (3.16)$$

With the same proof as above, one can easily see that Theorem 3.4
holds without any change for such a u_ϵ .

3.3. $\underline{W^{2,p}(\Omega)\text{-Regularity}}$ $(2 \leq p < +\infty)$.

In this part, we shall assume that Γ the boundary of Ω and the a_{ij} are sufficiently smooth. More precisely, we shall assume that for any $2 \leq p < +\infty$ when $f \in L^p(\Omega)$, the solution of the Dirichlet problem

$$\begin{cases} - \dfrac{\partial}{\partial x_i} (a_{ij} \dfrac{\partial u}{\partial x_j}) = f \\ u \in H_0^1(\Omega) \end{cases} \tag{3.17}$$

is in $W^{2,p}(\Omega)$ and that we have the estimate

$$||u||_{2,p} \leq C_p |f|_p \tag{3.18}$$

(see [4], [19]) where C_p is a constant which doesn't depend on u or f. Now let ϕ, ψ be two functions in $H^1(\Omega)$ such that

$$\phi(x) \leq \psi(x) \quad \text{a.e. in} \quad \Omega \tag{3.19}$$

and such that

$$\phi \leq 0 \leq \psi \quad \text{on} \quad \Gamma. \tag{3.20}$$

(3.20) is taken in the trace sense (see [82]) or simply in the usual sense if ϕ, ψ are smooth enough. Moreover, assume that $A\phi$ and $A\psi$ are measures such that

$$(-A\phi)^+, \ (-A\psi)^- \in L^p(\Omega). \tag{3.21}$$

Then if β_1, β_2 are two smooth functions (for instance C^1) satisfying (3.9) and (3.10) and if u_ε, for $\varepsilon > 0$, is the solution of (see 3.11)

$$\begin{cases} -Au_\varepsilon + \dfrac{\beta_1}{\varepsilon}(u_\varepsilon - \phi) + \dfrac{\beta_2}{\varepsilon}(u_\varepsilon - \psi) = f \\ u_\varepsilon \in H_0^1(\Omega), \end{cases} \tag{3.22}$$

then we have:

<u>Theorem 3.8</u>. Under the above assumptions (3.18)-(3,20) and for $f \in L^p(\Omega)$, the functions $\dfrac{\beta_1}{\varepsilon}(u_\varepsilon - \phi)$ and $\dfrac{\beta_2}{\varepsilon}(u_\varepsilon - \psi)$ are in $L^p(\Omega)$ and

$$\left|\dfrac{\beta_1}{\varepsilon}(u_\varepsilon - \phi)\right|_p \leq \left|f - (-A\phi)^+\right|_p \tag{3.23}$$

$$\left|\dfrac{\beta_2}{\varepsilon}(u_\varepsilon - \psi)\right|_p \leq \left|f + (-A\psi)^-\right|_p. \tag{3.24}$$

<u>Proof</u>: (See [29]). For $F(t) = |t|^{p-2}t$, let us consider

$$v = F(\frac{\beta_1}{\varepsilon}(u_\varepsilon - \phi)).$$

As we will see, it is no restriction to assume that $\frac{\beta_1}{\varepsilon}(u_\varepsilon - \phi)$ is
bounded. So by Theorem 2.3 and the assumptions on ϕ and β_1 ((3.20),
(3.9)), v is in $H_0^1(\Omega)$, and from (3.22), we deduce that

$$<-A(u_\varepsilon-\phi),v> \ + \ <-A\phi,v> \ + \int_\Omega (\frac{\beta_1}{\varepsilon}(u_\varepsilon-\phi) \ + \frac{\beta_2}{\varepsilon}(u_\varepsilon-\psi))\cdot vdx = \int_\Omega fvdx. \quad (3.25)$$

Now, since the function $\gamma = F \circ \frac{\beta_1}{\varepsilon}$ is monotone, we get

$$<-A(u_\varepsilon-\phi),v> \ = \ <-A(u_\varepsilon-\phi),\gamma(u_\varepsilon-\phi)>$$

$$= \int_\Omega \gamma'(u_\varepsilon-\phi)a_{ij} \frac{\partial}{\partial x_i}(u_\varepsilon-\phi)\frac{\partial}{\partial x_j}(u_\varepsilon-\phi)dx \geq 0. \quad (3.26)$$

Moreover, β_1 is negative, so is v, and

$$<-A\phi,v> \ \geq \int_\Omega (-A\phi)^+\cdot v \ dx. \quad (3.27)$$

Finally, we note that $v \neq 0$ only when $u_\varepsilon \leq \phi \leq \psi$ and hence when
$\frac{\beta_2}{\varepsilon}(u_\varepsilon-\psi) = 0$. From this it follows that

$$\int_\Omega (\frac{\beta_1}{\varepsilon}(u_\varepsilon-\phi) \ + \frac{\beta_2}{\varepsilon}(u_\varepsilon-\psi))\cdot v \ dx$$

$$= \int_\Omega \frac{\beta_1}{\varepsilon}(u_\varepsilon-\phi)\cdot vdx = |\frac{\beta_1}{\varepsilon}(u_\varepsilon-\phi)|_p^p. \quad (3.28)$$

Combining (3.25)-(3.28), we get

$$|\frac{\beta_1}{\varepsilon}(u_\varepsilon-\phi)|_p^p \leq \int_\Omega (f - (-A\phi)^+)\cdot v.$$

By applying Hölder's Inequality, we get with $q = p/p-1$

$$|\frac{\beta_1}{\varepsilon}(u_\varepsilon-\phi)|_p^p \leq |f - (-A\phi)^+|_p|v|_q.$$

But recalling that $v = |\frac{\beta_1}{\varepsilon}(u_\varepsilon-\phi)|^{p-2}\cdot \frac{\beta_1}{\varepsilon}(u_\varepsilon-\phi)$, we have

$$|v|_q = |\frac{\beta_1}{\varepsilon}(u_\varepsilon - \phi)|_p^{p-1}$$

and (3.23) follows. Now if $\frac{\beta_1}{\varepsilon}(u_\varepsilon-\phi)$ is not bounded, it is clear that
the truncated function $F_n(t) = F(t)$ if $|t| \leq n$, $F_n(t) = $ Sign $t\cdot F(n)$
if $|t| > n$, leads to a $v = F_n(\frac{\beta_1}{\varepsilon}(u_\varepsilon-\phi))$ which is in $H_0^1(\Omega)$. The result

then follows after an easy passage to the limit in n. Finally, to obtain (3.24), it is enough to consider $v = F(\frac{\beta_2}{\varepsilon}(u_\varepsilon - \psi))$ (or $v = F_n(\frac{\beta_2}{\varepsilon}(u_\varepsilon - \psi))$ which concludes the proof.

Remark 3.9. Assuming that $f \in L^\infty(\Omega), \phi, \psi \in W^{2,\infty}(\Omega)$ and taking the limit as $p \to +\infty$ in (3.23), (3.24) leads to the fact that $\frac{\beta_1}{\varepsilon}(u_\varepsilon - \phi), \frac{\beta_2}{\varepsilon}(u_\varepsilon - \psi) \in L^\infty(\Omega)$ with

$$|\frac{\beta_1}{\varepsilon}(u_\varepsilon - \phi)|_\infty \le |f - (-A\phi)^+|_\infty \qquad |\frac{\beta_2}{\varepsilon}(u_\varepsilon - \psi)|_\infty \le |f + (-A\psi)^-|_\infty. \qquad (3.29)$$

For the obstacle problems, we claim now that under the assumption (3.18), we have:

Theorem 3.10: Let ϕ be in $H^1(\Omega)$ with $\phi \le 0$ on Γ. If $A\phi$ is a measure such that $(-A\phi)^+ \in L^p(\Omega)$ and if $f \in L^p(\Omega)$ $(p \ge 2)$, then the solution u_1 of

$$u_1 \in K_\phi, \quad <-Au_1, v - u_1> \ge <f, v - u_1> \quad \forall v \in K_\phi \qquad (3.30)$$

is in $W^{2,p}(\Omega)$ and there exists a constant C_p independent of u_1, f, ϕ such that

$$||u_1||_{2,p} \le C_p(|f|_p + |(-A\phi)^+|_p). \qquad (3.31)$$

Theorem 3.11. Let ϕ, ψ be in $H^1(\Omega)$ and satisfy (3.19), (3.20) and (3.21). Then for $f \in L^p(\Omega)$ the solution u_2 of the problem

$$u_2 \in K_\phi^\psi, \quad <-Au_2, v - u_2> \ge <f, v - u_2> \quad \forall v \in K_\phi^\psi \qquad (3.32)$$

is in $W^{2,p}(\Omega)$ and there exists a constant C_p independent of u_2, f, ϕ, ψ such that

$$||u_2||_{2,p} \le C_p(|f|_p + |(-A\phi)^+|_p + |(-A\psi)^-|_p). \qquad (3.33)$$

Proof: First, for the proof of Theorem 3.10, consider u_ε the solution of (3.22) with $\beta_2 \equiv 0$ and $\beta_1(t) < 0$ $\forall t < 0$. Of course, for this u_ε (3.23) is satisfied and from (3.22), we get

$$|Au_\varepsilon|_p \le |f|_p + |f - (-A\phi)^+|_p.$$

By (3.18), we get

$$||u_\varepsilon||_{2,p} \le C_p(|f|_p + |(-A\phi)^+|_p) \qquad (3.34)$$

where C_p doesn't depend on ε. So we can extract a subsequence u_ε which converges weakly to some u in $W^{2,p}(\Omega)$ when $\varepsilon \to 0$. Now by Theorem 3.4, u is necessarily equal to u_1 which is in $W^{2,p}(\Omega)$ and (3.30) results from the lower semicontinuity of the norm in a Banach space. To prove Theorem 3.11, it is enough to consider the solution u_ε of (3.22) with $\beta_1(t) < 0$, $\forall t < 0$, $\beta_2(t) > 0$ $\forall t > 0$. From (3.18), (3.22), (3.23), and (3.24), we get

$$||u_\varepsilon||_{2,p} \leq C_p(|f|_p + |(-A\phi)^+|_p + |(-A\psi)^-|_p). \tag{3.35}$$

Since $u_\varepsilon \to u_2$ when $\varepsilon \to 0$ (by Theorem 3.4), the result follows as above.

Remark 3.12. When the a_{ij} are smooth, the above assumptions are satisfied for $f \in L^p(\Omega), \phi, \psi \in W^{2,p}(\Omega)$. (See [28], p. 8 for further comments.)

Remark 3.13. If $p > n$, by the Sobolev embedding theorem, we get that u_1 and u_2 are in $C^{1,\alpha}(\overline{\Omega})$ with $\alpha = 1 - \frac{n}{p}$. Note that in general $u_i \notin C^2(\Omega)$ (see below). For optimal results of regularity see [29].

Remark 3.14. Estimates (3.34), (3.35) could be used in place of (3.15) to go to the limit in (3.22).

3.4.   Some Complementary Results

In the unidimensional case, the solution of (3.30) or (3.32) is always continuous since $H_0^1(\Omega) \subset C(\overline{\Omega})$. In higher dimension, under the assumptions on the previous part, we can prove:

Theorem 3.15. Assume that $\phi \in C(\overline{\Omega})$ and $K_\phi \neq \emptyset$. If $f \in H^{-1}(\Omega)$ is such that the solution u of (3.17) is continuous on $\overline{\Omega}$, then the solution u_1 of

$$u_1 \in K_\phi, \quad <-Au_1, v - u_1> \geq <f, v - u_1> \quad \forall v \in K_\phi \tag{3.36}$$

is continuous in $\overline{\Omega}$.

Proof: (See [84].) Assume first that $f = 0$ and choose a sequence of smooth functions ϕ^n such that $\phi^n \leq \phi$ and $\phi^n \to \phi$ in $L^\infty(\Omega)$. Let u^n be the sequence defined by

$$u^n \in K_{\phi^n}, \quad <-Au^n, v - u^n> \geq 0 \quad \forall v \in K_{\phi^n}.$$

By Theorem 3.10, we have $u^n \in C(\overline{\Omega})$ and by the Proposition 3.2

$$|u^n - u_1|_\infty \leq |\phi^n - \phi|_\infty.$$

So $u^n \to u_1$ uniformly and u_1 is continuous.

Now to relate the case $f = 0$ to the general one, we use the following device due to Brezis. Let u be the solution of (3.17) i.e.,

$$u \in H_0^1(\Omega), \quad -Au = f \text{ in } \Omega.$$

We see from (3.36) that $u_1 - u$ satisfies

$$(u_1-u) \in K_\phi - u = K_{\phi-u}, \quad <-A(u_1-u),(v-u) - (u_1-u)> \geq 0 \quad \forall(v-u) \in K_{\phi-u}$$

and by the result in the case $f = 0$ we have that $u_1 - u$ is continuous in $\overline{\Omega}$ and since u is continuous in $\overline{\Omega}$ so is u_1.

Remark 3.16. By usualy regularity results, 3.15 holds, in particular for $f \in W^{-1,p}(\Omega)$ with $p > n$, Ω being smooth enough.

Under the assumptions of Theorem 3.15, we can divide Ω in two sets

$$0 = \{x \in \Omega | u_1(x) > \phi(x)\} \tag{3.37}$$

$$I = \{x \in \Omega | u_1(x) = \phi(x)\}. \tag{3.38}$$

By Theorem 3.15, 0 is open and I is closed. I is called "the coincidence set" and its boundary is the free boundary of the obstacle problem. The interested reader will find a study of this free boundary and also an extended definition of (3.37), and (3.38) in [76].

Note that in Ω, we have

$$-Au_1 \geq f \tag{3.39}$$

in the sense of distributions and so in the sense of measures. Indeed, for $\zeta \in \mathscr{D}(\Omega)$, $\zeta \geq 0$, choose $v = u_1 + \zeta$ in (3.36), we get

$$<-Au_1,\zeta> \geq <f,\zeta> \quad \zeta \in \mathscr{D}(\Omega), \quad \zeta \geq 0$$

and so (3.39) holds.

Moreover, we have

$$-Au_1 = f \text{ in } 0, \tag{3.40}$$

Indeed, for $\zeta \in \mathscr{D}(0)$, $\zeta \geq 0$ and $\varepsilon > 0$ small enough $v = u_1 - \varepsilon\zeta$ is a test function for (3.36). So, we get

$$<-Au_1,-\varepsilon\zeta> \geq <f,-\varepsilon\zeta> \quad \forall\zeta \in \mathscr{D}(0), \quad \zeta \geq 0$$

$$<-Au_1,\zeta> \leq <f,\zeta> \quad \forall\zeta \in \mathscr{D}(0), \quad \zeta \geq 0$$

and this inequality with (3.39) leads to (3.40) and so for smooth data, u_1 is the solution of the nonlinear problem

$$u_1 \in H_0^1(\Omega), \quad -Au_1 \geq f, \quad u_1 \geq \phi, \quad (-Au_1-f)\cdot(u_1-\phi) = 0, \tag{3.41}$$

that is to say:

$$u_1 \in H_0^1(\Omega) \qquad \mathrm{Max}(f + Au_1, \; \phi - u_1) = 0. \tag{3.42}$$

(See [33] to compare with the Bellman-Dirichlet problem.)

We now use this characterization of u_1 to complete our Remark 3.13, that is to say, to prove that u_1 is not generally in $C^2(\Omega)$ even if the data are smooth. Consider the one dimensional case with $\Omega = (-1,1)$, $A = d^2/dx^2$, $f = 0$ and $\phi(x) = -x^2 + \frac{3}{4}$ (see the figure below).

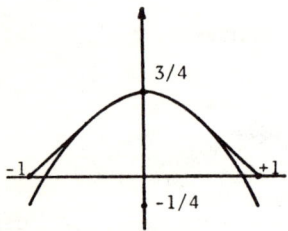

First note that the solution will be symmetric and we have to find it for instance on $(0,1)$. From (3.40), we have $u_1'' = 0$, and hence $u_1 = ax + b$, on $u_1 > \phi$. Since $u_1(1) = 0$, this gives

$$u_1 = a(1 - x)$$

on the component of $u_1 > \phi$ near 1. The fact that u_1 touches ϕ at a point x_0 can be expressed by

$$u_1(x_0) = \phi(x_0), \quad u_1'(x_0) = \phi'(x_0).$$

($u_1 - \phi$ achieves its minimum at x_0) which leads to consider u_1 defined by

$$u_1(x) = \begin{cases} x + 1 & \text{if } x \in [-1,-\frac{1}{2}] \\ \phi(x) & \text{if } x \in [-\frac{1}{2},+\frac{1}{2}]. \\ -x + 1 & \text{if } x \in [\frac{1}{2},1] \end{cases} \tag{3.43}$$

Now it is easy to check that u_1 is a solution of

$$u_1 \in K_\phi, \quad \int_{-1}^{+1} u_1'(v' - u_1')\,dx \geq 0 \qquad \forall v \in K_\phi. \tag{3.44}$$

Indeed, from (3.43), we deduce that

$$
u_1' = \begin{cases}
1 & \text{if } x \in (-1, -\tfrac{1}{2}) \\
-2x & \text{if } x \in (-\tfrac{1}{2}, \tfrac{1}{2}) \\
-1 & \text{if } x \in (\tfrac{1}{2}, 1)
\end{cases}
\tag{3.45}
$$

and so for all smooth v in K_ϕ, we get

$$
\int_{-1}^{+1} u_1' \cdot (v' - u_1')\,dx = \int_{-1}^{-1/2} (v' - u_1')\,dx + \int_{-1/2}^{1/2} -2x(v-u_1)'\,dx + \int_{1/2}^{1} -(v'-u_1')\,dx
$$

$$
= (v-u_1)(-\tfrac{1}{2}) + \int_{-1/2}^{1/2} 2(v-u_1)\,dx - (v-u_1)(\tfrac{1}{2})
$$

$$
- (v-u_1)(-\tfrac{1}{2}) + (v-u_1)(\tfrac{1}{2})
$$

$$
= \int_{-1/2}^{1/2} 2(v-\phi)\,dx \geq 0,
$$

and thus u_1 satisfies (3.44). Now, it is clear from (3.45) that u_1 is not C^2. However, from (3.45), we deduce that $u_1'' \in L^\infty(\Omega)$, and as we will see now this property is always true for sufficiently smooth data and has an extension to more than one dimension.

3.5. $W^{2,\infty}(\Omega)$-Regularity

As we mentioned in the preceding section, we now give a proof that for smooth data the second derivatives of solutions of obstacles problems are bounded up to the boundary.

To be precise, we will assume here that

$$
a_{ij} \in C^4(\overline{\Omega}), \quad \Gamma \text{ is of class } C^4.
\tag{3.46}
$$

For $\phi, \psi \in W^{2,\infty}(\Omega)$ satisfying (3.19), (3.20) and for $\varepsilon > 0$ we will first consider u_ε the solution of

$$
\begin{cases}
-Au_\varepsilon + \dfrac{\beta_1}{\varepsilon}(u_\varepsilon - \phi) + \varepsilon(u_\varepsilon - \phi) + \dfrac{\beta_2}{\varepsilon}(u_\varepsilon - \psi) + \varepsilon(u_\varepsilon - \psi) = 0 \\
u_\varepsilon \in H_0^1(\Omega).
\end{cases}
\tag{3.47}
$$

With β_1, β_2 two Lipschitz functions of class C^2 satisfying

$$
\beta_1(t) = 0 \quad \forall t \geq 0; \quad \beta_1'(t) \geq 0, \quad \beta_1''(t) \leq 0 \quad \forall t
\tag{3.48}
$$

$$
\beta_2(t) = 0 \quad \forall t \leq 0; \quad \beta_2'(t) \geq 0, \quad \beta_2''(t) \geq 0 \quad \forall t
\tag{3.49}
$$

The advantage of (3.47) over (3.11) (see Remark 3.7) is that if we put

$$\alpha_i(t) = \frac{\beta_i}{\epsilon}(t) + \epsilon t, \quad i = 1,2 \tag{3.50}$$

we have

$$\alpha_i'(t) \geq \epsilon > 0 \quad \forall t. \tag{3.51}$$

This will be useful in our estimates. (One could avoid this "ϵt" by using a weak maximum principle.)

Now (3.47) can be rewritten as

$$\begin{cases} Au_\epsilon = \alpha_1(u_\epsilon - \phi) + \alpha_2(u_\epsilon - \psi) \\ u_\epsilon \in H_0^1(\Omega). \end{cases} \tag{3.52}$$

Our first goal will be to obtain bounds for the second derivatives of u_ϵ which are independent of $\epsilon \in (0,1]$. Then, an argument similar to the proofs of Theorems 3.10, and 3.11 will provide us with $W^{2,\infty}$-regularity for obstacle problems.

The estimates of u_ϵ will be performed in three steps, first we will bound Au_ϵ, then the second derivatives of u_ϵ on the boundary Γ of Ω and finally inside Ω. So let us consider first:

3.5.1. A Uniform Bound for Au_ϵ.

We first note that $Au_\epsilon \in L^\infty(\Omega)$. Indeed, this follows immediately from (3.47) and the fact that $u_\epsilon \in L^\infty(\Omega)$ and satisfies

$$|u_\epsilon|_\infty \leq \text{Max}(|\phi|_\infty, |\psi|_\infty). \tag{3.53}$$

To prove this last inequality we put $k = \text{Max}(|\phi|_\infty, |\psi|_\infty)$. Since $(u_\epsilon - k)^+ \in H_0^1(\Omega)$ and by (3.52) we get

$$<-Au_\epsilon, (u_\epsilon - k)^+> = -\int_\Omega [\alpha_1(u_\epsilon - \phi) + \alpha_2(u_\epsilon - \psi)](u_\epsilon - k)^+ dx \leq 0,$$

since when $u_\epsilon \geq k$ we have $u_\epsilon \geq \phi$, $u_\epsilon \geq \psi$ and so $\alpha_1(u_\epsilon - \phi) + \alpha_2(u_\epsilon - \psi) \geq 0$. Now we deduce from the above inequality (see Proposition 3.2) that $u_\epsilon \leq k$. The inequality $u_\epsilon \geq -k$ follows in a similar way. We thus have 3.53 and $Au_\epsilon \in L^\infty(\Omega)$. Moreover, we have:

Proposition 3.17. If $\phi, \psi \in W^{2,\infty}(\Omega)$ satisfy (3.19) and (3.20), then the solution u_ϵ of (3.47) is such that $Au_\epsilon \in L^\infty(\Omega)$ and satisfies the estimate:

$$|Au_\epsilon|_\infty \leq \text{Max}[|(-A\phi)^+|_\infty, |(-A\psi)^-|_\infty] + 8\epsilon \text{Max}(|\phi|_\infty, |\psi|_\infty). \tag{3.54}$$

Proof: Applying the Remark 3.9 with $f = \varepsilon(u_\varepsilon - \phi) + \varepsilon(u_\varepsilon - \psi)$, we obtain, since $\beta_1 \leq 0$ and $\beta_2 \geq 0$

$$|\frac{\beta_1}{\varepsilon}(u_\varepsilon - \phi) + \frac{\beta_2}{\varepsilon}(u_\varepsilon - \psi)|_\infty \leq \text{Max}[|f - (-A\phi)^+|_\infty, |f + (-A\psi)^-|_\infty]$$

$$\leq \text{Max}[|(-A\phi)^+|_\infty, |(-A\psi)^-|_\infty] + |f|_\infty.$$

Now the result follows from (3.47) since by (3.53) we have $|f|_\infty \leq 4\varepsilon\text{Max}(|\phi|_\infty, |\psi|_\infty)$.

Remark 3.18. It results from (3.54) that under the assumptions of Proposition 3.17, for $\varepsilon \in (0,1]$:

$$|Au_\varepsilon|_\infty \leq C(||\phi||_{2,\infty} + ||\psi||_{2,\infty}) \tag{3.55}$$

where C depends on the a_{ij}'s but not on ε. Moreover, by (3.55) (see (3.18) we have $u_\varepsilon \in W^{2,p}(\Omega)$ and

$$||u_\varepsilon||_{2,p} \leq C_p|Au_\varepsilon|_p \leq C_p(||\phi||_{2,\infty} + ||\psi||_{2,\infty})$$

for all $p \geq 2$. Using the continuity of the embedding of $W^{2,p}(\Omega)$ in $C^1(\overline{\Omega})$ for $p > n$ we deduce that there is some constant C depending only on Ω and the a_{ij}'s such that

$$||u_\varepsilon||_{1,\infty} \leq C(||\phi||_{2,\infty} + ||\psi||_{2,\infty}). \tag{3.56}$$

(See also the Section 3.6 for an improvement of this.)

Remark 3.19. Taking the limit when $\varepsilon \to 0$, (3.54) provides us with

$$|Au_1|_\infty \leq |(-A\phi)^+|_\infty, \quad |Au_2|_\infty \leq \text{Max}[|(-A\phi)^+|_\infty, |(-A\psi)^-|_\infty]$$

where u_1 and u_2 are the solutions of (3.30), (3.32) with $f = 0$.

Remark 3.20. In the case where $n = 1$, it is clear that (3.55) already provides us with the estimate

$$||u_\varepsilon||_{2,\infty} \leq C(||\phi||_{2,\infty} + ||\psi||_{2,\infty})$$

and so in the sequel we shall assume that $n \geq 2$.

3.5.2. Boundary Estimate

First let us note that since $u_\varepsilon \in W^{2,p}(\Omega)$ (see Remark 3.18) we deduce from (3.46), (3.47) that $u_\varepsilon \in W^{4,p}(\Omega)$ $(p > n)$. This will allow us in the sequel to apply the maximum principle on the second derivatives of u_ε in the sense of Bony [25]. (One could choose at first smooth data and use an approximation argument.)

Now u_ε is at least in $C^2(\overline{\Omega})$ so the following proposition makes sense:

<u>Proposition 3.21</u>. Under the assumptions of the Proposition 3.17, if u_ε denotes the solution of (3.47) there exists a constant C which doesn't depend on $\varepsilon \in (0,1]$ or on ϕ,ψ such that for all $i,j = 1,\ldots,n$ we have

$$\left| \frac{\partial^2 u_\varepsilon}{\partial x_i \partial x_j} (x) \right| \leq C(||\phi||_{2,\infty} + ||\psi||_{2,\infty}) \qquad \forall x \in \Gamma. \qquad (3.57)$$

<u>Proof</u>: We will use here ideas from H. Brezis - D. Kinderlehrer [34] and R. Jensen [69].

Step 1: Let $x_0 \in \Gamma$. We will first show that there is no loss of generality in assuming that Γ is flat at x_0 and that the a_{ij}'s satisfy some additional properties. (The arguments below are classical in boundary value problems and to follow the idea of the proof instead of technical tools the reader is invited to go directly to (3.67) and (3.68).)

Since Γ is smooth, there exists a diffeomorphism $\theta = (\theta_1,\theta_2,\ldots,\theta_n)$ from a neighborhood V of x_0 in $\mathbb{R}^n$ onto B an open ball of $\mathbb{R}^n$ centered at 0 such that $\theta(x_0) = 0$ and

$$\theta(\Omega \cap V) = B^+ = \{X = (X_1,\ldots,X_n) \in B \,|\, X_n > 0\}$$

$$\theta(\Gamma \cap V) = \partial_0 B^+ = \{X = (X_1,\ldots,X_n) \in B \,|\, X_n = 0\}.$$

For each function v defined on $V \cap \overline{\Omega}$, we define a function v' on $B^+ \cup \partial_0 B^+$ (and conversely) by setting

$$v'(X) = v(\theta^{-1}(X)) \iff v(x) = v'(\theta(x)). \qquad (3.58)$$

So we have (setting u in place of u_ε)

$$\frac{\partial u}{\partial x_i}(x) = \sum_{k=1}^n \frac{\partial u'}{\partial X_k} (\theta(x)) \cdot \frac{\partial \theta_k}{\partial x_i}(x) \qquad (3.59)$$

$$\frac{\partial^2 u}{\partial x_i \partial x_j}(x) = \sum_{k,\ell=1}^n \frac{\partial^2 u'}{\partial X_\ell \partial X_k}(\theta(x)) \cdot \frac{\partial \theta_\ell}{\partial x_j}(x) \cdot \frac{\partial \theta_k}{x_i}(x)$$

$$+ \sum_{k=1}^n \frac{\partial u'}{\partial X_k}(\theta(x)) \cdot \frac{\partial^2 \theta_k}{\partial x_i \partial x_j}(x). \qquad (3.60)$$

Since $\dfrac{\partial u'}{\partial X_i}$ and $\dfrac{\partial^2 u'}{\partial X_i \partial X_j}$ are given by a similar formula it results from (3.56) that the derivatives $\dfrac{\partial u'}{\partial X_i}$ are bounded on B^+ by the quantity

$C(||\phi||_{2,\infty} + ||\psi||_{2,\infty})$. Moreover, from this bound and (3.59) (3,60) we easily see that to have (3.57) in a neighborhood of x_0 in $\Gamma \cap V$ it is enough for $\dfrac{\partial^2 u'}{\partial X_i \partial X_j}$ to satisfy an inequality of type (3.57) with ϕ',ψ' in place of ϕ,ψ on a neighborhood of 0 in $\partial_0 B^+$. Thus, via this trans-formation θ we have to study u' satisfying (see (3.52), (3.59), (3.60)) with the summation convention:

$$
\begin{cases}
a'_{ij}\dfrac{\partial^2 u'}{\partial X_i \partial X_j} + b'_k \dfrac{\partial u'}{\partial X_k} = \alpha_1(u'-\phi') + \alpha_2(u'-\psi') & \text{on } B^+ \\[2mm]
u' = 0 & \text{on } \partial_0 B^+
\end{cases}
\tag{3.61}
$$

the coefficients a'_{ij} being given by (see (3.60))

$$
a'_{ij}(X) = \sum_{k,\ell=1}^{n} \left(a_{k\ell}\frac{\partial\theta_i}{\partial x_k}\frac{\partial\theta_j}{\partial x_\ell}\right)(\theta^{-1}(X)).
\tag{3.62}
$$

Replacing a'_{ij} by $(a'_{ij} + a'_{ji})/2$ it is clear that there is no loss of generality in assuming that in (3.61) we have

$$
a'_{ij} = a'_{ji} \qquad \forall i,j = 1,\ldots,n.
\tag{3.63}
$$

(Note also that from (3.62) we can deduce that (3.1) holds for a'_{ij} with some ν' in place of ν.) Now let us perform a second transformation $T = (T_1,\ldots,T_n)$ defined by (see [61], [69]).

$$
T_i(X) = X_i + X_n\phi_i(\overline{X}) \qquad X = (X_1,\ldots,X_n) = (\overline{X},X_n), \text{ if } i \neq n
$$

$$
T_n(X) = X_n
$$

where ϕ_i is given by

$$
\phi_i(\overline{X}) = -a'_{ni}(\overline{X},0)/a'_{nn}(\overline{X},0).
\tag{3.64}
$$

An easy computation shows that the Jacobian of T at 0 is 1 and so in a neighborhood of 0, T is an isomorphism which preserves $\partial_0 B^+$ and maps B^+ into B^+. Because of this second transformation, and with ob-vious notations for functions v'' (see (3.58)), we are now concerned with the study (in some ball λB^+ $(\lambda < 1)$ still denoted by B^+) of u'' the solution of

$$
\begin{cases}
a''_{ij}\dfrac{\partial^2 u''}{\partial X_i \partial X_j} + b''_k \dfrac{\partial u''}{\partial X_k} = \alpha_1(u''-\phi'') + \alpha_2(u''-\psi'') & \text{on } B^+ \\[2mm]
u'' = 0 & \text{on } \partial_0 B^+
\end{cases}
\tag{3.65}
$$

with (see (3.62), (3.63)) $a''_{ij} = a''_{ji}$, but now (see (3.62), (3.64)) for all
$i = 1, \ldots, n-1$

$$a''_{in} = a''_{ni} = \sum_{k,\ell=1}^{n} a'_{k\ell} \cdot \frac{\partial T_n}{\partial X_k} \cdot \frac{\partial T_i}{\partial X_\ell} = a'_{ni} + a'_{nn} \phi_i = 0 \quad \text{on} \quad \partial_0 B^+.$$

Rewriting the first equality in (3.65) as

$$\frac{\partial}{\partial X_i}(a''_{ij} \frac{\partial u''}{\partial X_j}) + (b''_k - \frac{\partial a''_{ik}}{\partial X_i}) \frac{\partial u''}{\partial X_k} = \alpha_1(u'' - \phi'') + \alpha_2(u'' - \psi'')$$

and letting w be the solution of

$$\begin{cases} - \frac{\partial}{\partial X_i}(a''_{ij} \frac{\partial w}{\partial X_j}) = (b''_k - \frac{\partial a''_{ik}}{\partial X_i}) \frac{\partial u''}{\partial X_k} \quad \text{on} \quad B^+ \\[2mm] w = 0 \quad \text{on} \quad \partial B^+, \end{cases} \qquad (3.66)$$

we see by setting $u^* = u'' - w$, $\phi^* = \phi'' - w$, $\psi^* = \psi'' - w$ that we have

$$\begin{cases} \frac{\partial}{\partial X_i}(a''_{ij} \frac{\partial u^*}{\partial X_j}) = \alpha_1(u^* - \phi^*) + \alpha_2(u^* - \psi^*) \quad \text{on} \quad B^+ \\[2mm] u^* = 0 \quad \text{on} \quad \partial_0 B^+. \end{cases}$$

Now from (3.56), (3.59), (3.60) it follows that the right-hand side of
the first equation in (3.66) is bounded in $W^{1,p}(B^+)$, for all p, by the
quantity $C(||\phi||_{2,\infty} + ||\psi||_{2,\infty})$, and so on B^+ or on some λB^+ $(\lambda < 1)$
we have $w \in W^{3,p} \subset W^{2,\infty}$ if $p > n$ and an estimation of type

$$||w||_{2,\infty} \leq C(||\phi||_{2,\infty} + ||\psi||_{2,\infty}).$$

Thus to prove (3.57) we have only to prove it on a neighborhood of 0 in
$\partial_0 B^+$ and with u^*, ϕ^*, ψ^* in place of u, ϕ, ψ. Reverting to our notation
in u, we have proved that in place of (3.52) we can assume that $u = u_\epsilon$
satisfies

$$\begin{cases} \frac{\partial}{\partial X_i} (a_{ij} \frac{\partial u}{\partial x_i}) = \alpha_1(u - \phi) + \alpha_2(u - \psi) \quad \text{on} \quad B^+ \\[2mm] u = 0 \quad \text{on} \quad \partial_0 B^+ \end{cases} \qquad (3.67)$$

with $a_{ij} = a_{ji}$ and

$$a_{in} = a_{ni} = 0 \quad \text{on} \quad \partial_0 B^+ \qquad \forall i = 1, \ldots, n-1, \qquad (3.68)$$

ϕ, ψ being in $W^{2,\infty}(B^+)$ satisfying always (3.19), (3.20) and (3.55),
(3.56) now being true on B^+.

Remark 3.22. The fact that (3.55) is preserved by the different transformation results from (3.56), (3.59), (3.60), (3.62). Note also that
the fact $u \in W^{4,p}$ is preserved since by (3.46), (3.64), $T \circ \theta$ is of class
C^4.

Step 2: We deal now with u satisfying (3.67), and (3.68) holding.
First since $u = 0$ on $\partial_0 B^+$ we have:

$$\frac{\partial^2 u}{\partial x_i \partial x_j} = 0 \quad \text{on } \partial_0 B^+ \qquad \forall i,j = 1,\ldots,n-1.$$

Using the first equation in (3.67) and (3.68) we get (see (3.19), (3.48),
(3.49), (3.50))

$$a_{nn}(x) \cdot \frac{\partial^2 u}{\partial x_n^2}(x) + \frac{\partial a_{nn}}{\partial x_n}(x) \cdot \frac{\partial u}{\partial x_n}(x) = -\varepsilon(\phi(x) + \psi(x)) \quad \forall x \in \partial_0 B^+. \quad (3.69)$$

And thus by the ellipticity of A (i.e., $a_{nn} \geq \nu > 0$) and from (3.56)
we get for $\varepsilon \in (0,1]$

$$\left| \frac{\partial^2 u}{\partial x_n^2}(x) \right| \leq C(||\phi||_{2,\infty} + ||\psi||_{2,\infty}) \qquad \forall x \in \partial_0 B^+ \qquad (3.70)$$

where C doesn't depend on ε. Thus it remains only to bound the derivatives of type $\partial^2 u/\partial x_k \partial x_n$, $k = 1,\ldots,n-1$ in a neighborhood of 0 in
$\partial_0 B^+$.

Step 3: We thus want to estimate $\partial^2 u/\partial x_k \partial x_n$. Before entering into
the technical argument let us briefly outline the proof. Assume that
$u = u_\varepsilon$ is a smooth solution of:

$$\begin{cases} \Delta u = \alpha_1(u - \phi) + \alpha_2(u - \psi) & \text{on } \mathbb{R}^n_+ \\ u = 0 & \text{on } x_n = 0 \end{cases} \qquad (3.71)$$

with $\mathbb{R}^n_+$ equal to the half space $\{(x_1,x_2,\ldots,x_n)|x_n > 0\}$, (3.55) and
(3.56) being satisfied in $\mathbb{R}^n_+$ with $A = \Delta$, α_1,α_2 as above, and $\phi \leq 0 \leq \psi$
on $x_n = 0$. Let us set

$$Bu = \frac{\partial^2 u}{\partial x_k \partial x_n} + \frac{\partial u}{\partial x_k}.$$

To get

$$\frac{\partial^2 u}{\partial x_k \partial x_n}(x) \leq C(||\phi||_{2,\infty} + ||\psi||_{2,\infty}) \qquad (3.72)$$

it is clearly sufficient (by (3.55) and (3.56)) to bound

$$A_+ u = Bu + \Delta u \quad \text{or} \quad A_- u = Bu - \Delta u$$

by $C(||\phi||_{2,\infty} + ||\psi||_{2,\infty})$. ($\pm$ gives us two chances!) Thus denoting by x^+ (Resp. x^-) a point of $\bar{\mathbb{R}}_+^n$ where $A_+ u$ (Resp. $A_- u$) achieves its maximum it is enough to prove that we have

$$A_+ u(x^+) \quad \text{or} \quad A_- u(x^-) \leq C(||\phi||_{2,\infty} + ||\psi||_{2,\infty}). \tag{3.73}$$

Two cases are possible

Case 1: $x^+ \in \{x_n = 0\}$.   Then of course in this case we have

$$\frac{\partial}{\partial x_n}(A_+ u)(x^+) \leq 0 \Longleftrightarrow \left(\frac{\partial^3 u}{\partial x_k \partial x_n^2} + \frac{\partial^2 u}{\partial x_k \partial x_n} + \frac{\partial}{\partial x_n} \Delta u\right)(x^+) \leq 0. \tag{3.74}$$

But on $x_n = 0$ we have (since $u = 0$, $\phi \leq 0 \leq \psi$, and by (3.48), (3.49))

$$\Delta u = \frac{\partial^2 u}{\partial x_n^2} = -\varepsilon(\phi + \psi).$$

Thus taking the derivative in the x_k direction

$$\frac{\partial^3 u}{\partial x_k \partial x_n^2}(x^+) = -\varepsilon \frac{\partial}{\partial x_k}(\phi + \psi)(x^+),$$

Using (3.71) we also easily obtain

$$\frac{\partial}{\partial x_n} \Delta u(x^+) = \varepsilon \frac{\partial}{\partial x_n}(2u - \phi - \psi)(x^+).$$

By (3.74), (3.56) we have now

$$\frac{\partial^2 u}{\partial x_k \partial x_n}(x^+) \leq C(||\phi||_{2,\infty} + ||\psi||_{2,\infty})$$

with C independent of $\varepsilon \in (0,1]$ and by (3.55), (3.56) this clearly leads to (3.73). The case where $x^- \in \{x_n = 0\}$ follows in the same way, so we can assume:

Case 2: x^+ and $x^- \in R_+^n$.   Then applying A_+ and A_- to the first equation of (3.71) we get:

$$\Delta(A_+ u) = \alpha_1'(u-\phi)(A_+ u - A_+ \phi) + \alpha_1'(u-\psi)(A_+ u - A_+ \psi)$$
$$+ \alpha_1''(u-\phi)Q_+(u-\phi) + \alpha_2''(u-\psi)Q_+(u-\psi) \tag{3.75}$$

$$\Delta(A_-u) = \alpha_1'(u-\phi)(A_-u-A_-\phi) + \alpha_1'(u-\psi)(A_-u-A_-\psi)$$

$$+ \alpha_1''(u-\phi)Q_-(u-\phi) + \alpha_2''(u-\psi)Q_-(u-\psi) \qquad (3.76)$$

where we have set

$$Q_\pm(v) = \frac{\partial v}{\partial x_k} \cdot \frac{\partial v}{\partial x_n} \pm |\nabla v|^2 \quad \text{(clearly } Q_+ \text{ is positive, } Q_- \text{ is negative)}.$$

By the maximum principle we have in this case

$$\Delta(A_+u)(x^+) \le 0 \quad \text{and} \quad \Delta(A_-u)(x^-) \le 0. \qquad (3.77)$$

If $x^+ \in [u \ge \phi] = \{x \in \mathbb{R}_+^n | u(x) \ge \phi(x)\}$ we have (see (3.48), (3.49))

$$\alpha_1''(u - \phi)(x^+) = 0 \quad \text{and} \quad \alpha_2''(u - \psi)(x^+) \ge 0.$$

Thus from (3.75), (3.77) we deduce (since $Q_+ \ge 0$)

$$\alpha_1'(u - \phi)(A_+u - A_+\phi) + \alpha_2'(u - \psi)(A_+u - A_+\psi)(x^+) \le 0$$

and by (3.51) we get

$$A_+u(x^+) \le A_+\phi(x+) \quad \text{or} \quad A_+\psi(x^+)$$

which clearly leads to (3.73). If $x^- \in [u \le \psi]$ we have

$$\alpha_1''(u - \phi)(x^-) \le 0 \quad \text{and} \quad \alpha_2''(u - \psi)(x^-) = 0$$

and since in this case $Q^- \le 0$, we get now as above

$$A_-u(x^-) \le A_-\phi(x^-) \quad \text{or} \quad A_-\psi(x^-)$$

which leads again to (3.73). So the only remaining case is when

$$x^+ \in [u < \phi] \quad \text{and} \quad x^- \in [u > \psi]. \qquad (3.78)$$

But this cannot occur. We have indeed from the definition of x^+, x^-

$$Bu + \Delta u(x^+) = A_+u(x^+) \ge A_+u(x^-) = Bu + \Delta u(x^-)$$

$$Bu - \Delta u(x^-) = A_-u(x^-) \ge A_-u(x^+) = Bu - \Delta u(x^+).$$

Adding these inequalities, we get

$$[\alpha_1(u-\phi) + \alpha_2(u-\psi)](x^+) = \Delta u(x^+) \ge \Delta u(x^-)$$

$$= [\alpha_1(u-\phi) + \alpha_2(u-\psi)](x^-).$$

But if (3.78) holds the above left hand side is strictly negative (see
(3.48), (3.49)) and the right side is strictly positive. This is a contradiction and hence (3.72) holds in all cases. The inequality in the
other side follows in the same way; note for instance that -u is a solution of a problem of the same type as (3.71) with $-\phi, -\psi$ in place of
ϕ, ψ.

Let us now return to (3.67).

With an eye to deriving the analogs of (3.75), (3.76), set

$$Bu = \frac{\partial^2 u}{\partial x_k \partial x_n} + c \frac{\partial u}{\partial x_k}, \quad k = 1, \ldots, n-1$$

where c is a positive constant satisfying

$$c - \frac{1}{a_{nn}} \cdot \frac{\partial a_{nn}}{\partial x_n} \geq 1. \tag{3.79}$$

Applying B to Au we easily obtain

$$B(Au) = A(Bu) + \frac{\partial}{\partial x_i}\left(B(a_{ij})\frac{\partial u}{\partial x_j} + \frac{\partial a_{ij}}{\partial x_k} \cdot \frac{\partial^2 u}{\partial x_n \partial x_j}\right.$$
$$\left. + \frac{\partial a_{ij}}{\partial x_n} \cdot \frac{\partial^2 u}{\partial x_k \partial x_j}\right). \tag{3.80}$$

Now, let σ be a smooth function such that

$$0 \leq \sigma \leq 1 \text{ on } B^+ \qquad \sigma = 1 \text{ in a neighborhood of } 0$$
$$\sigma = 0 \text{ on } \partial B^+ - \partial_0 B^+ \qquad \frac{\partial \sigma}{\partial x_n} = 0 \text{ on } \partial_0 B^+. \tag{3.81}$$

(One of the roles of this function is to avoid the problems created by the
fact that u is not prescribed on a part of the boundary of B^+.)

Setting for i = 1,...,n (with the summation convention in j)

$$S_i = B(a_{ij})\frac{\partial u}{\partial x_j} + \frac{\partial a_{ij}}{\partial x_k} \cdot \frac{\partial^2 u}{\partial x_n \partial x_j} + \frac{\partial a_{ij}}{\partial x_n} \cdot \frac{\partial^2 u}{\partial x_k \partial x_j}$$

and multiplying (3.80) by σ, we get after an easy computation:

$$\sigma B(Au) = A(\sigma Bu) + (A\sigma \cdot Bu - \frac{\partial \sigma}{\partial x_i} \cdot S_i) + \frac{\partial}{\partial x_i}(\sigma S_i - 2a_{ij}\frac{\partial \sigma}{\partial x_j} Bu)$$
$$= A(\sigma Bu) + f_0 + \frac{\partial f_i}{\partial x_i} \tag{3.82}$$

with obvious definition for the f_i's.

Now let us temporarily assume that we have proved the following lemma:

Lemma 3. If the f_i denote the functions of (3.82), there exists a w in $W^{2,p}(B^+)$ $(p > n)$ which is a unique solution of the problem

$$
\begin{cases}
-Aw = f_0 + \dfrac{\partial f_i}{\partial x_i} & \text{on } B^+ \\[2mm]
w = 0 \text{ on } \partial B^+ - \partial_0 B^+, \quad \dfrac{\partial w}{\partial x_n} = 0 \text{ on } \partial_0 B^+.
\end{cases}
\tag{3.83}
$$

Moreover, there exists a constant C which doesn't depend on $\varepsilon \in (0,1]$, ϕ, ψ such that:

$$||w||_\infty \le C(||\phi||_{2,\infty} + ||\psi||_{2,\infty}).
\tag{3.84}$$

Now from (3.82) we deduce

$$A(\sigma Bu - w) = \sigma B(Au).$$

Using (3.67) to compute $B(Au)$ we get

$$
A(\sigma Bu-w) = \alpha_1'(u-\phi)(\sigma Bu-\sigma B\phi) + \alpha_2'(u-\psi)(\sigma Bu-\sigma B\psi)
$$
$$
+ \alpha_1''(u-\phi)\sigma \frac{\partial}{\partial x_k}(u-\phi)\frac{\partial}{\partial x_n}(u-\phi) + \alpha_2''(u-\Psi)\sigma \frac{\partial}{\partial x_k}(u-\psi)\frac{\partial}{\partial x_n}(u-\psi).
\tag{3.85}
$$

On the other hand, applying A to Au given by (3.67) leads to

$$
A(Au) = \alpha_1'(u-\phi)(Au-A\phi) + \alpha_2'(u-\psi)(Au-A\psi)
$$
$$
+ \alpha_1''(u-\phi)a_{ij}\frac{\partial}{\partial x_i}(u-\phi)\frac{\partial}{\partial x_j}(u-\psi) + \alpha_2''(u-\psi)a_{ij}\frac{\partial}{\partial x_i}(u-\phi)\frac{\partial}{\partial x_j}(u-\psi).
\tag{3.86}
$$

Multiplying (3.85) by ν, adding and subtracting $A(Au)$ give us now the following equations to which we will apply the maximum principle:

$$
A(\nu(\sigma Bu-w) + Au) = \alpha_1'(u-\phi)(A_+u-A_+\phi) + \alpha_2'(u-\psi)(A_+u-A_+\psi)
$$
$$
+ \alpha_1''(u-\phi)Q_+(u-\phi) + \alpha_2''(u-\psi)Q_+(u-\psi)
$$
$$
A(\nu(\sigma Bu-w) - Au) = \alpha_1'(u-\phi)(A_-u-A_-\phi) + \alpha_2'(u-\psi)(A_-u-A_-\psi)
$$
$$
+ \alpha_1''(u-\phi)Q_-(u-\phi) + \alpha_2''(u-\psi)Q_-(u-\psi).
\tag{3.87}
$$

In these equalities we have set by definition:

$$A_\pm v = \nu\sigma Bv \pm Av$$

$$Q_\pm v = \nu\sigma \frac{\partial v}{\partial x_k} \cdot \frac{\partial v}{\partial x_n} \pm a_{ij}\frac{\partial v}{\partial x_i}\frac{\partial v}{\partial x_j}.$$

We shall note that by (3.1), (3.81) we have:

$$a_{ij} \frac{\partial v}{\partial x_i} \cdot \frac{\partial v}{\partial x_j} \geq \nu |\nabla v|^2 \geq \nu \sigma \left| \frac{\partial v}{\partial x_k} \right| \left| \frac{\partial v}{\partial x_n} \right|$$

and thus, Q_+ is positive and Q_- is negative.

Let us now denote by x^+ (Resp. x^-) a point where $\nu(\sigma Bu - w) + Au$ (Resp. $\nu(\sigma Bu - w) - Au$) achieves its maximum in $\overline{B^+}$. We want to prove that

$$\sigma \frac{\partial^2 u}{\partial x_k \partial x_n}(x) \leq C(||\phi||_{2,\infty} + ||\psi||_{2,\infty}) \qquad \forall x \in \overline{B^+}. \tag{3.89}$$

By the definition of x^+, x^-, it suffices to (see (3.55), (3.56), (3.84)) prove that

$$(\nu(\sigma Bu-w) + Au)(x^+) \quad \text{or} \quad (\nu(\sigma Bu-w) - Au)(x^-)$$
$$\leq C(||\phi||_{2,\infty} + ||\psi||_{2,\infty}). \tag{3.90}$$

Different cases are possible:

Case 0: $x^+ \in \partial B^+ - \partial_0 B^+$. We have then, by definition of x^+ and since in this case $\sigma(x^+) = 0$ and $w(x^+) = 0$:

$$(\nu(\sigma Bu - w) + Au)(x^+) = Au(x^+),$$

and by (3.55) we get (3.90). Clearly the case $x^- \in \partial B^+ - \partial_0 B^+$ follows in the same way.

Case 1: $x^+ \in \partial_0 B^+$ (see Case 1 of the heuristic proof). We have then (recall that $\nu(\sigma Bu-w) + Au \in W^{2,p}(B^+)$, $(p > n)$)

$$\frac{\partial}{\partial x_n}(\nu(\sigma Bu - w) + Au)(x^+) \leq 0.$$

Using (3.67), (3.81), (3.83), the above inequality can be rewritten as

$$\left[\nu\sigma \cdot \frac{\partial}{\partial x_n} Bu + \alpha_1'(u-\phi)\frac{\partial}{\partial x_n}(u-\phi) + \alpha_2'(u-\psi)\frac{\partial}{\partial x_n}(u-\psi) \right](x^+) \leq 0.$$

But on $\partial_0 B^+$ we have $u = 0$ and thus, $\alpha_1'(u-\phi)(x^+) = \alpha_2'(u-\psi)(x^+) = \varepsilon$. Combined with the definition of B this leads to

$$\left[\nu\sigma \frac{\partial^3 u}{\partial x_k \partial x_n^2} + \nu\sigma c \frac{\partial^2 u}{\partial x_k \partial x_n} + \varepsilon \frac{\partial}{\partial x_n}(2u - \phi - \psi) \right](x^+) \leq 0.$$

To estimate the third derivative of u in the above inequality we take the derivative of (3.69) in the x_k direction and we get

$$a_{nn} \frac{\partial^3 u}{\partial x_k \partial x_n^2} = - \frac{\partial a_{nn}}{\partial x_k} \frac{\partial^2 u}{\partial x_k \partial x_n} - \frac{\partial a_{nn}}{\partial x_k} \frac{\partial^2 u}{\partial x_n^2} - \frac{\partial^2 a_{nn}}{\partial x_k \partial x_n} \frac{\partial u}{\partial x_n} - \varepsilon \frac{\partial}{\partial x_k}(\phi + \psi).$$

Dividing by a_{nn}, which is greater or equal to ν, and replacing $\dfrac{\partial^3 u}{\partial x_k \partial x_n^2}$

by its value in our inequality above leads after using (3.56), (3.70) to

$$\nu\sigma\left(c - \frac{1}{a_{nn}} \frac{\partial a_{nn}}{\partial x_n}\right) \frac{\partial^2 u}{\partial x_k \partial x_n}(x^+) \leq C(||\phi||_{2,\infty} + ||\psi||_{2,\infty})$$

where C doesn't depend on $\varepsilon \in (0,1]$. But now from (3.79) we deduce that

$$\nu\sigma \frac{\partial^2 u}{\partial x_k \partial x_n}(x^+) \leq C(||\phi||_{2,\infty} + ||\psi||_{2,\infty}),$$

and (3.90) follows easily. Clearly the same argument holds if $x^- \in \partial_0 B^+$
and thus we can assume that we are in the:

Case 2. x^+ and x^- are in B^+.

Let us assume first that $x^+ \in [u \geq \phi]$. Then we have (see (3.48),
(3.49))

$$\alpha_1''(u - \phi)(x^+) = 0 \quad \text{and} \quad \alpha_2''(u - \psi)(x^+) \geq 0. \tag{3.91}$$

Now the maximum principle in $W^{2,p}(\Omega)$ (p > n) (see [25]) leads to

$$\text{ess·lim inf}_{x \to x^+} A(\nu(\sigma Bu - w) + Au)(x) \leq 0.$$

But from (3.91), (3.87) and the fact that Q_+ is positive we get

$$\text{ess·lim inf}_{x \to x^+}[\alpha_1'(u-\phi)(A_+u-A_+\phi) + \alpha_2'(u-\psi)(A_+u-A_+\psi)](x) \leq 0.$$

From (3.51) one easily deduces that this implies

$$A_+u(x^+) \leq |A_+\phi|_\infty \quad \text{or} \quad |A_+\psi|_\infty.$$

Using now (3.84) we get (3.90). The case $x^- \in [u \leq \psi]$ follows in
the same way, see Case 2 in the heuristic proof, and the case where
$x^+ \in [u < \phi]$ and $x^- \in [u > \psi]$ is impossible. To see this, just add the
two inequalities

$$\nu(\sigma Bu-w)(x^+) + Au(x^+) \geq \nu(\sigma Bu-w)(x^-) + Au(x^-)$$

$$\nu(\sigma Bu-w)(x^-) - Au(x^-) \geq \nu(\sigma Bu-w)(x^+) + Au(x^+)$$

to get a contradiction. (As after 3.78) thus (3.90) holds in all cases.

The lower bound can be obtained by noting that $-u$ satisfies a problem of the form (3.67) with now ϕ, ψ changed to $-\phi$, $-\psi$. By (3.89) the estimate (3.57) holds in a neighborhood of x_0 and a compactness argument shows that (3.57) is true globally. Thus it remains only to prove Lemma 3.

Step 4:   Proof of Lemma 3.

First we remark that $f_0 + \partial f_i / \partial x_i$ can be written as

$$g_0 + \frac{\partial g_i}{\partial x_i}$$

where the g_i's are of the same type as the f_i (i.e., in $W^{2,p}(B^+)$ and bounded in $L^p(B^+)$ by the quantity $C(||\phi||_{2,\infty} + ||\psi||_{2,\infty})$ -- see (3.82), (3.55) but also with

$$g_n(x) = 0 \quad \text{on} \quad \partial_0 B^+. \tag{3.92}$$

Clearly, it is enough to check that $\partial f_n / \partial x_n$ can be written like this. But (see the form of f_n in (3.82)) in $\partial f_n / \partial x_n$ only the terms

$$\frac{\partial}{\partial x_n}(\alpha \frac{\partial^2 u}{\partial x_n^2}) \quad \text{and} \quad \frac{\partial}{\partial x_n}(\beta \frac{\partial^2 u}{\partial x_i \partial x_j}), \quad j \neq n$$

may not be of the above type. To see that they can be suitably rewritten, note that we have:

$$\frac{\partial}{\partial x_n}(\alpha \frac{\partial^2 u}{\partial x_n^2}) = \frac{\partial}{\partial x_n} \alpha(\frac{\partial^2 u}{\partial x_n} + \frac{1}{a_{nn}}(\frac{\partial a_{nn}}{\partial x_n} \cdot \frac{\partial u}{\partial x_n} + \epsilon\phi + \epsilon\psi))$$

$$- \frac{\partial}{\partial x_n} \frac{\alpha}{a_{nn}}(\frac{\partial a_{nn}}{\partial x_n} \frac{\partial u}{\partial x_n} + \epsilon\phi + \epsilon\psi)$$

$$\frac{\partial}{\partial x_n}(\beta \frac{\partial^2 u}{\partial x_i \partial x_j}) = \frac{\partial}{\partial x_i} \frac{\partial}{\partial x_n}(\beta \frac{\partial u}{\partial x_j}) - \frac{\partial}{\partial x_n}(\frac{\partial \beta}{\partial x_i} \frac{\partial u}{\partial x_j})$$

as (3.69) allows us to conclude.

We then set for $x = (\bar{x}, x_n) \in B$, $\text{sign}(x_n)$ denoting the sign of x_n (i.e., ± 1),

$$\tilde{a}_{in}(x) = \tilde{a}_{ni}(x) = \text{sign}(x_n)a_{in}(\bar{x},|x_n|) \quad \forall i = 1,\ldots,n-1$$

$$\tilde{a}_{ij}(x) = \tilde{a}_{ji}(x) = a_{ij}(\bar{x},|x_n|) \quad \text{for all pair} \quad (i,j) \neq (k,n)$$

$$(k = 1,\ldots,n-1)$$

$$\tilde{g}_i(x) = g_i(\bar{x},|x_n|) \quad \forall i = 0,\ldots,n-1,$$

$$\tilde{g}_n(x) = \text{sign}(x_n) \cdot g_n(\bar{x},|x_n|)$$

Now it is easy to check that the $\tilde{a}_{ij}(x)$ satisfy the ellipticity condition (3.1) and thus, there exists an unique solution $\tilde{w} \in H^1(B)$ of the problem

$$\begin{cases} -\dfrac{\partial}{\partial x_i}(\tilde{a}_{ij}\dfrac{\partial \tilde{w}}{\partial x_j}) = \tilde{g}_0 + \dfrac{\partial}{\partial x_i}\tilde{g}_i \quad \text{on} \quad B \\[2ex] \tilde{w} = 0 \quad \text{on} \quad \Gamma. \end{cases}$$

The g_i's are in $W^{1,p}(B^+)$ and the $\tilde{g}_i$'s are in $W^{1,p}(B)$ (see (3.92)), thus $\tilde{g}_0 + \partial\tilde{g}_i/\partial x_i \in L^p(B)$ and (see [4], [65]) $\tilde{w} \in W^{2,p}(B) \subset C^1(\bar{B})$. Moreover, it is easy to check that $\tilde{w}(x,x_n) = \tilde{w}(x,-x_n)$ since these two functions are solutions of the above problem. Hence

$$\frac{\partial \tilde{w}}{\partial x_n}(x) = 0 \quad \text{on} \quad \partial_0 B^+$$

and the restriction of w to B^+ satisfies (3.83). The estimate (3.84) results simply from the theorem A_1 of the appendix combined with (3.55).

Remark 3.22. In the case where $\phi = \psi = 0$ on Γ this proposition is simpler to prove using an unpublished result of H. Brezis. (See [51].)

We can now conclude the estimates of the second derivatives of u_ε by:

3.5.3. Global Estimates.

Proposition 3.23. If $\phi,\psi \quad W^{2,\infty}(\Omega)$ satisfy (3.19), (3.20), and under the assumptions of Proposition 3.17, then the solution u_ε of (3.47) is in $W^{2,\infty}(\Omega)$ and there exists a constant C which doesn't depend on $\varepsilon \in (0,1], \phi, \psi$ such that

$$||u_\varepsilon||_{2,\infty} \leq C(||\phi||_{2,\infty} + ||\psi||_{2,\infty}). \tag{3.93}$$

Proof: First, let us estimate the derivative $\partial^2 u_\varepsilon/\partial x_k \partial x_\ell$. Set (with u in place of u_ε)

$$Bu = \frac{\partial^2 u}{\partial x_k \partial x_\ell}.$$

We have

$$B(Au) = A(Bu) + \frac{\partial}{\partial x_i}(B(a_{ij})\frac{\partial u}{\partial x_j} + \frac{\partial a_{ij}}{\partial x_k}\frac{\partial^2 u}{\partial x_j \partial x_\ell} + \frac{\partial a_{ij}}{\partial x_\ell}\frac{\partial^2 u}{\partial x_j \partial x_k}).$$

By introducing the solution w of

$$\begin{cases} -Aw = \dfrac{\partial}{\partial x_i}\left(B(a_{ij})\dfrac{\partial u}{\partial x_j} + \dfrac{\partial a_{ij}}{\partial x_k}\dfrac{\partial^2 u}{\partial x_j \partial x_\ell} + \dfrac{\partial a_{ij}}{\partial x_\ell}\dfrac{\partial^2 u}{\partial x_j \partial x_k}\right) \\[2ex] w = 0 \quad \text{on } \Gamma \end{cases}$$

the above inequality becomes

$$A(Bu - w) = B(Au)$$

with, moreover, (see the Theorem A_1 of the Appendix and (3.55))

$$||w||_\infty \le C(||\phi||_{2,\infty} + ||\psi||_{2,\infty}). \qquad (3.94)$$

Now computing $B(Au)$ by replacing Au by its value (see (3.52)) we get

$$A(Bu - w) = \alpha_1'(u-\phi)(Bu-B\phi) + \alpha_2'(u-\psi)(Bu-B\psi)$$

$$+ \alpha_1''(u-\phi)\frac{\partial}{\partial x_k}(u-\phi)\frac{\partial}{\partial x_\ell}(u-\phi) + \alpha_2''(u-\psi)\frac{\partial}{\partial x_k}(u-\psi)\frac{\partial}{\partial x_\ell}(u-\psi).$$

Multiplying the two sides of this equality by ν, then adding and subtracting (3.86) we get the following two equations:

$$A(\nu(Bu-w) + Au) = \alpha_1'(u-\phi)(A_+u-A_+\phi) + \alpha_2'(u-\psi)(A_+u-A_+\psi)$$

$$+ \alpha_1''(u-\phi)Q_+(u-\phi) + \alpha_2''(u-\psi)Q_+(u-\psi)$$

$$A(\nu(Bu-w) - Au) = \alpha_1'(u-\phi)(A_-u-A_-\phi) + \alpha_2'(u-\psi)(A_-u-A_-\psi)$$

$$+ \alpha_1''(u-\phi)Q_-(u-\phi) + \alpha_2''(u-\psi)Q_-(u-\psi)$$

where we have set

$$A_\pm v = \nu Bv \pm Av$$

$$Q_\pm v = \nu \frac{\partial v}{\partial x_k} \cdot \frac{\partial v}{\partial x_\ell} \pm a_{ij}\frac{\partial v}{\partial x_i}\frac{\partial v}{\partial x_j}.$$

(Note that $a_{ij}\dfrac{\partial v}{\partial x_i}\dfrac{\partial v}{\partial x_j} \pm \nu \dfrac{\partial v}{\partial x_k}\dfrac{\partial v}{\partial x_\ell} \ge \nu(|\nabla v|^2 \pm \dfrac{\partial v}{\partial x_k}\dfrac{\partial v}{\partial x_\ell}) \ge 0$ and $Q_\pm$ is of the sign $\pm$.)

The proof is now as in the preceding proposition. That is to say, we denote by x^+ (Resp. x^-) a point in $\overline{\Omega}$ where $\nu(Bu - w) + Au$ (Resp. $\nu(Bu - w) - Au$) achieves its maximum. If x^+ (or x^-) belongs to Γ, then by (3.55), (3.57) and the definition of x^+ we have

$$\nu(Bu-w)(x) + Au(x) \le \nu(Bu-w)(x^+) + Au(x^+) \le C(||\phi||_{2,\infty} + ||\psi||_{2,\infty}),$$

$$\forall x \in \overline{\Omega}.$$

Applying (3.55) and (3.94) we get

$$|Bu(x)| \leq C(||\phi||_{2,\infty} + ||\psi||_{2,\infty}).\qquad\qquad (3.95)$$

Thus we can assume $x^+, x^- \in \Omega$, but applying verbatim (with $B = \Omega$, $\sigma = 1$) the arguments of Case 2 of the previous proposition leads again to (3.95) which completes the proof (for a variant of this technique, see [52]).

As a consequence for the obstacle problems we now have:

Theorem 3.24. (See [34] and [69]). Under the assumptions (3.46) if $\phi \in W^{2,\infty}(\Omega)$, $\phi \leq 0$ on Γ, $f \in W^{1,p}(\Omega)$, $p > n$ then the solution u_1 of the problem

$$\begin{cases} <-Au_1, v - u_1> \geq <f, v - u_1> \quad \forall v \in K_\phi \\ u_1 \in K_\phi = \{v \in H_0^1(\Omega) \,|\, v(x) \geq \phi(x) \quad a.e. \ in \ \Omega\} \end{cases}$$

is in $W^{2,\infty}(\Omega)$ and there exists a constant C which doesn't depend on u_1, ϕ, f such that

$$||u_1||_{2,\infty} \leq C(||f||_{1,p} + ||\phi||_{2,\infty}).\qquad\qquad (3.96)$$

Theorem 3.25. (See [51].) Under the assumptions (3.46) if $\phi, \psi \in W^{2,\infty}(\Omega)$, $\phi \leq 0 \leq \psi$ on Γ, $f \in W^{1,p}(\Omega)$, $p > n$, then the solution u_2 of the problem

$$\begin{cases} <-Au_2, v - u_2> \geq <f, v - u_2> \quad \forall v \in K_\phi^\psi \\ u_2 \in K_\phi^\psi = \{v \in H_0^1(\Omega) \,|\, \psi(x) \geq v(x) \geq \phi(x) \quad a.e. \ in \ \Omega\} \end{cases}$$

is in $W^{2,\infty}(\Omega)$ and there exists a constant C which doesn't depend on u_2, ϕ, ψ, f such that

$$||u_2||_{2,\infty} \leq C(||f||_{1,p} + ||\phi||_{2,\infty} + ||\psi||_{2,\infty}).\qquad\qquad (3.97)$$

Proof: Let u_ε be the solution of (3.16). By introducing the solution w of

$$-Aw = f, \qquad w \in H_0^1(\Omega)$$

we easily see that $u_\varepsilon - w$ satisfies (3.47) with $\phi - w$, $\psi - w$ in place of ϕ, ψ. Thus from the Theorem 3.23 we get that $u_\varepsilon - w \in W^{2,\infty}(\Omega)$ and by (3.93)

$$||u_\varepsilon - w||_{2,\infty} \leq C(||\phi - w||_{2,\infty} + ||\psi - w||_{2,\infty}).$$

By regularity results for the Dirichlet problem (see [4]) and the Sobolev embedding theorem (see [1])

$$||w||_{2,\infty} \leq C||w||_{3,p} \leq C'||f||_{1,p}$$

and the above inequality leads now to

$$||u_\varepsilon||_{2,\infty} \leq C(||f||_{1,p} + ||\phi||_{2,\infty} + ||\psi||_{2,\infty}). \qquad (3.98)$$

Under the assumptions (3.48), (3.49) taking

$$\beta_1(t) < 0 \qquad \forall t < 0, \quad \beta_2(t) \equiv 0, \quad \psi = ||\phi||_{2,\infty}$$

then we have that (see Proposition 3.4 and Remark 3.7) $u_\varepsilon \to u_1$ as $\varepsilon \to 0$ and a standard argument gives us Theorem 3.24, (3.96) resulting from (3.98). Taking

$$\beta_1(t) < 0 \qquad \forall t < 0, \quad \beta_2(t) > 0 \qquad \forall t > 0$$

leads to Theorem 3.25.

3.6. $W^{1,\infty}(\Omega)$-Regularity

First let us take a look at the case of an operator with constant coefficients. In this case the translation method works and we are even able to get $C^{0,\alpha}(\bar{\Omega})$ regularity for all $0 < \alpha \leq 1$. Indeed, assuming here that Ω is a bounded Lipschitz domain of $\mathbb{R}^n$ we have:

<u>Theorem 3.26.</u> Let ϕ, ψ be two functions in $H_0^1(\Omega) \cap C^{0,\alpha}(\bar{\Omega})$ $(0 < \alpha \leq 1)$ with $\phi \leq \psi$. Assume that μ is a constant and u is the solution of

$$u \in K_\phi^\psi, \quad \int_\Omega \nabla u \cdot \nabla(v-u)\,dx \geq \int_\Omega \mu \cdot (v-u)\,dx \quad \forall v \in K_\phi^\psi. \qquad (3.99)$$

Then $u \in C^{0,\alpha}(\bar{\Omega})$ and we have

$$|u|_\alpha \leq \text{Max}(|\phi|_\alpha, |\psi|_\alpha)$$

where

$$|v|_\alpha = \sup_{\substack{x \neq y \\ (x,y) \in \Omega}} \frac{|v(x) - v(y)|}{|x - y|^\alpha}.$$

<u>Proof</u>: (See H. Brezis - N. Sibony [36] and P. Hartman - G. Stampacchia [68].) For $v \in H_0^1(\Omega)$ let us denote by $\tilde{v}$ the extension of v by 0 outside Ω. for $v \in C^{0,\alpha}(\bar{\Omega})$ we have clearly

$$|\tilde{v}(x + h) - \tilde{v}(x)| \leq |v|_\alpha |h|^\alpha \quad \forall x,h \in \mathbb{R}^n. \qquad (3.100)$$

Now fix $h \in \mathbb{R}^n$ and consider for $C = \text{Max}(|\phi|_\alpha, |\psi|_\alpha)$ the functions

$$v(x) = \text{Max}(\tilde{u}(x), \tilde{u}(x+h) - C|h|^{\alpha}), \quad w(x) = \text{Min}(\tilde{u}(x), \tilde{u}(x-h) + C|h|^{\alpha}).$$

By (3.100) we have in $\mathbb{R}^n$

$$\tilde{\phi}(x) \leq \tilde{u}(x) \leq v(x) \leq \text{Max}(\tilde{\psi}(x), \tilde{\psi}(x+h) - C|h|^{\alpha}) = \tilde{\psi}(x)$$

$$\tilde{\phi}(x) \leq \text{Min}(\tilde{\phi}(x), \tilde{\phi}(x-h) + C|h|^{\alpha}) \leq w(x) \leq \tilde{u}(x) \leq \tilde{\psi}(x)$$

and so the restrictions to Ω of $v(x) = \tilde{u}(x) + (\tilde{u}(x+h) - \tilde{u}(x) - C|h|^{\alpha})^{+}$
and $w(x) = \tilde{u}(x) - (\tilde{u}(x-h) - \tilde{u}(x) + C|h|^{\alpha})^{-}$ are in K_{ϕ}^{ψ}. Applying (3.99)
we get

$$\int_{\mathbb{R}^n} \nabla\tilde{u}(x)\cdot\nabla(\tilde{u}(x+h) - \tilde{u}(x) - C|h|^{\alpha})^{+}dx \geq \int_{\mathbb{R}^n} \mu(\tilde{u}(x+h) - \tilde{u}(x) - C|h|^{\alpha})^{+}dx$$

$$\int_{\mathbb{R}^n} \nabla\tilde{u}(x)\cdot\nabla-(\tilde{u}(x-h) - \tilde{u}(x) + C|h|^{\alpha})^{-}dx \geq \int_{\mathbb{R}^n} -\mu(\tilde{u}(x-h) - \tilde{u}(x) + C|h|^{\alpha})^{-}dx.$$

Now changing x into $x + h$ in the integrals of the second inequality
gives us

$$\int_{\mathbb{R}^n} \nabla\tilde{u}(x+h)\cdot\nabla-(\tilde{u}(x) - \tilde{u}(x+h) + C|h|^{\alpha})^{-}dx$$

$$\geq -\int_{\mathbb{R}^n} \mu(\tilde{u}(x) - \tilde{u}(x+h) + C|h|^{\alpha})^{-}dx.$$

Thus adding with the first one, using that $(-f)^{-} = f^{+}$, leads to

$$\int_{\mathbb{R}^n} \nabla(\tilde{u}(x) - \tilde{u}(x+h))\cdot\nabla(\tilde{u}(x+h) - \tilde{u}(x) - C|h|^{\alpha})^{+}dx \geq 0$$

$$\Longleftrightarrow \int_{\mathbb{R}^n} |\nabla(\tilde{u}(x+h) - \tilde{u}(x) - C|h|^{\alpha})^{+}|^{2} \leq 0 \qquad (\forall h \in \mathbb{R}^n).$$

So for almost all x we have

$$\tilde{u}(x + h) - \tilde{u}(x) \leq C|h|^{\alpha} \quad \forall h \in \mathbb{R}^n.$$

That is to say, after possibly changing $\tilde{u}$ on a set of measure 0

$$\tilde{u}(x) - \tilde{u}(y) \leq C|x - y|^{\alpha}, \qquad \forall x,y \in \mathbb{R}^n.$$

Now a permutation of x and y gives the result, i.e.,

$$|u(x) - u(y)| \leq C|x - y|^{\alpha}, \qquad \forall x,y \in \overline{\Omega}.$$

Remark 3.27. The above method works also, of course, for inequalities of
type

$$u \in K_{\phi}^{\psi}, \quad <-Au,v - u> \geq <f,v - u> \quad \forall v \in K_{\phi}^{\psi}$$

where   A   is an operator with constant coefficients and   f   is in a suitable space.   (Indeed, if Ω is smooth enough one can easily reduce the problem to the case $f = 0$.)

We now choose $f = \mu$ to give a well-known application of this result to the elastic-plastic torsion problem.   Let us denote by δ the Euclidean distance to the boundary, i.e.,

$$\delta(x) = \inf_{y \in \Gamma} |x - y| \qquad \forall x \in \Omega$$

with $|\;|$ for the usual Euclidean norm in $\mathbb{R}^n$.

Then under the previous assumptions upon Ω we have:

__Theorem 3.28.__   If μ is a positive constant the problems

$$\begin{cases} u \in K = \{v \in H_0^1(\Omega) \,|\, v(x) \leq \delta(x) \quad \text{a.e.} \quad \text{in} \quad \Omega\} \\ \displaystyle\int_\Omega \nabla u \cdot \nabla (v-u)\, dx \geq \int_\Omega \mu(v-u)\, dx \qquad \forall v \in K \end{cases} \tag{3.101}$$

and

$$\begin{cases} u \in K' = \{v \in H_0^1(\Omega) \,|\, |\nabla v(x)| \leq 1 \quad \text{a.e.} \quad \text{in} \quad \Omega\} \\ \displaystyle\int_\Omega \nabla u \cdot \nabla (v-u)\, dx \geq \int_\Omega \mu(v-u)\, dx \qquad \forall v \in K' \end{cases} \tag{3.102}$$

have the same solution.

__Proof:__   See [36].   By uniqueness of the solution of both problems it is enough to prove that the solution   u   of (3.101) satisfies also (3.102). First take $v = u^+$ in (3.101) we get that $u \geq 0$ and so   K   can be replaced in (3.101) by

$$K^+ = \{v \in H_0^1(\Omega) \,|\, 0 \leq v(x) \leq \delta(x) \quad \text{a.e.} \quad \text{in} \quad \Omega\}.$$

But now $0, \delta$ are Lipschitz with Lipschitz constant less than   1, so Theorem 3.26 tells us that   u   is in   K'   and the inclusion $K' \subset K$ shows that   u   satisfies (3.102).   One can again remark that the theorem holds if the above operator $A = \Delta$ is replaced by an elliptic one with constant coefficients.

__Remark 3.29.__   One can show also that the following equalities hold (see [77], [42])

$$\{x \in \Omega \,|\, u(x) < \delta(x)\} = \{x \in \Omega \,|\, |\nabla u(x)| < 1\}$$

$$\{x \in \Omega \,|\, u(x) = \delta(x)\} = \{x \in \Omega \,|\, |\nabla u(x)| = 1\}.$$

For applications of Theorem 3.28 see [77], [42]. Note also that some
extensions can be given to the non simply connected case (i.e., in the
case of (2.33)) see [42], [43], [53].

Let us now consider the case where A has variable coefficients.

We shall assume in this case that Ω is a smooth bounded open set
of $\mathbb{R}^n$ (that is to say at least of class C^1) and that Ω satisfies the
following exterior sphere condition (see [55]):

> There exists $R > 0$ such that for all $x_0 \in \Gamma$ there
> exists a ball of radius R which meets $\overline{\Omega}$ only at x_0.   (3.103)

Remark 3.30. In the case $n = 1$ clearly (3.103) holds. For $n \geq 2$ one
can see easily that this holds for Ω of class C^2.

The coefficients of A will be chosen such that

$$a_{ij} \in W^{1,\infty}(\Omega) \tag{3.104}$$

and as previously, for $\phi, \psi \in W^{1,\infty}(\Omega)$ satisfying (3.19), (3.20) and for
$\varepsilon > 0$ we shall first consider u_ε the solution of the problem

$$\begin{cases} Au_\varepsilon = \dfrac{\beta_1}{\varepsilon}(u_\varepsilon - \phi) + \dfrac{\beta_2}{\varepsilon}(u_\varepsilon - \psi) \\[2mm] u_\varepsilon \in H_0^1(\Omega) \end{cases} \tag{3.105}$$

where β_1 and β_2 are two smooth functions satisfying (3.9), (3.10).

In what follows we shall assume that the a_{ij}'s are in $C^1(\overline{\Omega})$. Since
the different estimates depend on the a_{ij} only through the norms
$||a_{ij}||_{1,\infty}$, it is clear that the results extend under the assumption (3.104).
Using now the same arguments as in 3.5.1 we have clearly (3.53) and thus
$Au_\varepsilon \in L^\infty(\Omega)$. But the usual regularity results for elliptic problems insure us that $u_\varepsilon \in W^{2,p}(\Omega) \subset C^1(\overline{\Omega})$ if $p > n$ and so the following proposition makes sense;

Proposition 3.31. For $\phi, \psi \in W^{1,\infty}(\Omega)$ and under the above assumptions
there exists a constant C which depends only on Ω, ν, R and the norms
$||a_{ij}||_{1,\infty}$ such that:

$$|\nabla u_\varepsilon(x)| \leq C \operatorname{Max}(||\nabla\phi^+||_\infty, ||\nabla\psi^-||_\infty), \qquad \forall x \in \Gamma. \tag{3.106}$$

Proof: Let $x_0 \in \Gamma$ and choose our coordinates such that 0 coincides
with the center of the ball of radius R which meets $\overline{\Omega}$ only at x_0.
Set, for x in Ω,

$$\delta^+(x) = k\left(\frac{1}{R^p} - \frac{1}{|x|^p}\right), \qquad \delta^-(x) = -\delta^+(x)$$

k,p being positive constants which will be chosen later on, $|x|$ denoting the usual Euclidean norm of x.

First we have

$$\delta^-(x_0) = u_\varepsilon(x_0) = \delta^+(x_0)$$
$$\delta^-(x) \le u_\varepsilon(x) \le \delta^+(x), \qquad \forall x \in \Gamma. \tag{3.107}$$

Moreover, an easy computation gives for x in Ω

$$\frac{\partial\delta^+(x)}{\partial x_i} = kpx_i|x|^{-(p+2)} \tag{3.108}$$

$$\frac{\partial^2\delta^+}{\partial x_i\partial x_j}(x) = kp\delta_{ij}|x|^{-(p+2)} - kp(p+2)x_ix_j|x|^{-(p+4)} \tag{3.109}$$

(with $\delta_{ij} = 1$ if $i = j, 0$ otherwise).

Since Ω is bounded, there exists a real number D such that Ω is included in the ball of center 0 and radius D. So assume first that k,p are chosen such that:

$$|\nabla\delta^+(x)| = kp|x|^{-(p+1)} \ge kpD^{-(p+1)} = \mathrm{Max}(||\nabla\phi^+||_\infty, ||\nabla\psi^-||_\infty) \tag{3.110}$$
$$\forall x \in \Omega.$$

Thus, if $x \in \Omega$ and if y denotes the first point where the segment $(x,0)$ meets Γ we have (note that δ^+ is radial)

$$\delta^+(x) \ge \delta^+(x) - \delta^+(y) \ge \mathrm{Max}(||\nabla\phi^+||_\infty, ||\nabla\psi^-||_\infty)\cdot|x-y| \ge \quad \begin{array}{l}\phi^+(x)-\phi^+(y)=\phi^+(x)\\[4pt]\psi^-(x)-\psi^-(y)=\psi^-(x).\end{array}$$

Hence if (3.110) holds we have in Ω

$$\phi \le \phi^+ \le \delta^+ \quad \text{and} \quad \delta^- \le -\psi^- \le \psi. \tag{3.111}$$

Now using (3.108), (3.109) we get

$$-A\delta^+(x) = -\sum_{i,j=1}^{n}\left(a_{ij}(x)\frac{\partial^2\delta^+(x)}{\partial x_i\partial x_j} + \frac{\partial}{\partial x_i}a_{ij}(x)\frac{\partial\delta^+}{\partial x_j}(x)\right)$$

$$= kp|x|^{-(p+2)}\cdot\sum_{i,j=1}^{n}\left[(p+2)a_{ij}(x)x_ix_j|x|^{-2} - \delta_{ij}a_{ij}(x) - \frac{\partial}{\partial x_j}a_{ij}(x)x_j\right]$$

Hence from (3.1) we deduce

$$-A\delta^+(x) \ge kp|x|^{-(p+2)}[\nu(p+2) - M] \tag{3.112}$$

where M is some constant which bounds from above on Ω the quantity

$$| \sum_{i,j=1}^{n} [\delta_{ij} a_{ij}(x) + \frac{\partial a_{ij}}{\partial x_i}(x) x_j] | .$$

Assuming M is greater than 2ν, we can clearly choose k,p such that (see (3.110))

$$\nu(p+2) - M = 0, \quad kpD^{-(p+1)} = \text{Max}(||\nabla\phi^+||_\infty, ||\nabla\psi^-||_\infty). \tag{3.113}$$

From (3.111), (3.112) we then deduce in Ω (see (3.9), (3.10))

$$-A\delta^+ + \frac{\beta_1}{\varepsilon}(\delta^+ - \phi) + \frac{\beta_2}{\varepsilon}(\delta^+ - \psi) \geq -A\delta^+ \geq 0 = -Au_\varepsilon + \frac{\beta_1}{\varepsilon}(u_\varepsilon - \phi) + \frac{\beta_2}{\varepsilon}(u_\varepsilon - \psi)$$

$$-A\delta^- + \frac{\beta_1}{\varepsilon}(\delta^- - \phi) + \frac{\beta_2}{\varepsilon}(\delta^- - \psi) \leq -A\delta^- \leq 0 = -Au_\varepsilon + \frac{\beta_1}{\varepsilon}(u_\varepsilon - \phi) + \frac{\beta_2}{\varepsilon}(u_\varepsilon - \psi).$$

Hence using (2.15), (2.16) and test functions $(u_\varepsilon - \delta^+)^+$, $(u_\varepsilon - \delta^-)^-$ one deduces from these inequalities that

$$\delta^-(x) \leq u_\varepsilon(x) \leq \delta^+(x) \qquad \forall x \in \Omega$$

and thus

$$|\nabla u_\varepsilon(x_0)| \leq |\nabla\delta^+(x_0)| = kpR^{-(p+1)}.$$

By (3.113) this finishes the proof.

Let us now deduce global estimates for u_ε:

Proposition 3.32. Under the assumptions of the previous proposition and if u_ε is the solution of (3.105) there exists a constant C which depends only on Ω, ν, R and the norms $||a_{ij}||_{1,\infty}$ (in particular not on ε) such that:

$$||\nabla u_\varepsilon||_\infty \leq C \cdot \text{Max}(||\nabla\phi||_\infty, ||\nabla\psi||_\infty). \tag{3.114}$$

Proof: First let us prove that we have

$$||\nabla u_\varepsilon||_2 \leq C \cdot \text{Max}(||\nabla\phi^+||_\infty, ||\nabla\psi^-||_\infty) \tag{3.115}$$

where C depends only on Ω, ν and the $||a_{ij}||_{1,\infty}$.
Indeed

$$v = \phi^+ - \psi^-$$

is a function in $H_0^1(\Omega)$ (see 3.20) such that

$$\phi \leq v \leq \psi.$$

(Since $\phi \leq \psi$ we have $\phi^+ \leq \psi^+$ and $\phi^- \geq \psi^-$, and so $\phi = \phi^+ - \phi^- \leq \phi^+ - \psi^- \leq \psi^+ - \psi^- = \psi$). Thus from (3.105) we get - since $\beta_1 = 0$ on $t \geq 0$ and $\beta_2 = 0$ on $t \leq 0$:

$$<-Au_\varepsilon, u_\varepsilon - v> = -\int_\Omega [\frac{\beta_1}{\varepsilon}(u_\varepsilon - \phi) + \frac{\beta_2}{\varepsilon}(u_\varepsilon - \psi) - \frac{\beta_1}{\varepsilon}(v-\phi) - \frac{\beta_2}{\varepsilon}(v-\psi)] \cdot (u_\varepsilon - v) dx \leq 0,$$

and so

$$\int_\Omega a_{ij}(x) \frac{\partial u_\varepsilon}{\partial x_i} \cdot \frac{\partial u_\varepsilon}{\partial x_j} dx \leq \int_\Omega a_{ij}(x) \frac{\partial u_\varepsilon}{\partial x_i} \cdot \frac{\partial v}{\partial x_j} dx.$$

Applying now (3.1) and the Cauchy-Schwarz inequality leads to

$$\nu ||\nabla u_\varepsilon||_2^2 \leq C ||\nabla u_\varepsilon||_2 ||\nabla v||_2 \tag{3.116}$$

which gives (3.114) by our choice of v.

Now let us take the derivative of the first equation of (3.105) in the x_k direction. We get (by writing u in place of u_ε)

$$-A(\frac{\partial u}{\partial x_k}) + \frac{\beta_1'}{\varepsilon}(u-\phi)(\frac{\partial u}{\partial x_k} - \frac{\partial \phi}{\partial x_k}) + \frac{\beta_2'}{\varepsilon}(u-\psi)(\frac{\partial u}{\partial x_k} - \frac{\partial \psi}{\partial x_k})$$

$$= \frac{\partial}{\partial x_i}(\frac{\partial a_{ij}}{\partial x_k}\frac{\partial u}{\partial x_j}). \tag{3.117}$$

Let us define T by

$$T = \text{Max}(C,1) \cdot \text{Max}(||\nabla \phi||_\infty, ||\nabla \psi||_\infty) \tag{3.118}$$

where C is the constant which occurs in (3.106) and Ω_1, Ω_2 the open sets (u_ε is in $W^{2,p}(\Omega)$):

$$\Omega_1 = \{x \in \Omega | \frac{\partial u}{\partial x_k} > T\}$$

$$\Omega_2 = \{x \in \Omega | \frac{\partial u}{\partial x_k} < -T\}.$$

Moreover, in Ω_1, Ω_2 respectively, let us set:

$$v_1 = \frac{\partial u}{\partial x_k} - T, \qquad v_2 = -\frac{\partial u}{\partial x_k} - T.$$

We thus obtain two positive functions in $H_0^1(\Omega_1)$ and $H_0^1(\Omega_2)$ which satisfy (see (3.117) and note that $\beta_i' \geq 0$)

$$-Av_1 = -A(\frac{\partial u}{\partial x_k} - T) = -A(\frac{\partial u}{\partial x_k}) \le \frac{\partial}{\partial x_i}(\frac{\partial a_{ij}}{\partial x_k} \cdot \frac{\partial u}{\partial x_j}) \quad \text{in} \quad \Omega_1$$

(3.119)

$$-Av_2 = -A(-\frac{\partial u}{\partial x_k} - T) = -A(-\frac{\partial u}{\partial x_k}) \le \frac{\partial}{\partial x_i}(-\frac{\partial a_{ij}}{\partial x_k} \cdot \frac{\partial u}{\partial x_j}) \quad \text{in} \quad \Omega_2.$$

But now we can apply the Corollary A_2 of the Appendix. More precisely, if $n = 2$, using (iii) of the Corollary A_2 from (3.115), (3.119) it results that $v_i \in L^q(\Omega_i)$ for all q $(i = 1,2)$ and that

$$|v_i|_{q,\Omega_i} \le C \, \text{Max}(||\nabla\phi||_\infty, ||\nabla\psi||_\infty).$$

($|\ \ |_{q,\Omega_i}$ denotes the L^q-norm in Ω_i and C some constant which doesn't depend on ε.) By dividing Ω into Ω_1, Ω_2 and $\{x \in \Omega | |\frac{\partial u}{\partial x_i}| \le T\}$ this provides us with $\frac{\partial u}{\partial x_k} \in L^q(\Omega)$ and an estimate

$$\left|\frac{\partial u}{\partial x_k}\right|_q \le C \, \text{Max}(||\nabla\phi||_\infty, ||\nabla\psi||_\infty),$$

for all $q \ge 2$. But now since this estimate holds clearly for all k, choosing $q > n$ we can apply the point (i) of Corollary A_2 to (3.119). Thus with the same decomposition of Ω as above we now get

$$\left|\frac{\partial u}{\partial x_k}\right|_\infty \le C \, \text{Max}(||\nabla\phi||_\infty, ||\nabla\psi||_\infty)$$

which is (3.114).

In the case $n = 1$, by (3.115) we are in case (i) of Corollary A_2 and one can conclude as above (of course, in this case one can provide a shorter proof). If now $n > 2$, (3.115) and (ii) of Corollary A_2 show that $v_i \in L^{2*}(\Omega_i)$ and an estimate

$$|v_i|_{2*,\Omega_i} \le C \, \text{Max}(||\nabla\phi||_\infty, ||\nabla\psi||_\infty).$$

But the decomposition of Ω into Ω_1, Ω_2 and $\{x \in \Omega | |\frac{\partial u}{\partial x_k}| \le T\}$ combined with the above inequality provide us now with:

$$\left|\frac{\partial u}{\partial x_k}\right|_{2*} \le C \cdot \text{Max}(||\nabla\phi||_\infty, ||\nabla\psi||_\infty).$$

Thus if $2* > n$ we can apply (i) of Corollary A_2 to conclude as previously. If $2* \le n$, and since the above inequality holds for all k, we get by applying the point (ii) of the Corollary A_2 with $p < 2*$:

$$\left|\frac{\partial u}{\partial x_k}\right|_{p*} \le C \cdot \text{Max}(||\nabla\phi||_\infty, ||\nabla\psi||_\infty)$$

for all $p < 2^*$. Repeating the procedure, it is clear that we can exceed n for some p^* and the result is now an application of (i) of Corollary A_2 which concludes the proof.

Remark 3.33. Because of (3.106) and the estimate (see (3.115))

$$||\nabla u_\varepsilon||_2 \le C \text{ Max}(||\nabla\phi^+||_2, ||\nabla\psi^-||_2)$$

one might expect that an estimate of type

$$||\nabla u_\varepsilon||_\infty \le C \text{ Max}(||\nabla\phi^+||_\infty, ||\nabla\psi^-||_\infty) \tag{3.120}$$

holds (this would lead to a similar estimate for the solution of the homogeneous obstacle problems). But this is not the case in general (see Remark 3.36).

Note that with the same proof as in (3.53) we have

$$|u_\varepsilon|_\infty \le \text{Max}(|\phi^+|_\infty, |\psi^-|_\infty).$$

As a consequence of the above estimates, we are now able to prove for the obstacle problems:

Theorem 3.34. Let $\phi \in W^{1,\infty}(\Omega)$, $\phi \le 0$ on Γ and $f \in L^p(\Omega)$ with $p > n$. If (3.17), (3.18) hold then under the above assumptions the solution u_1 of

$$u_1 \in K_\phi, \quad <-Au_1, v - u_1> \ge <f, v - u_1> \quad \forall v \in K_\phi$$

is in $W^{1,\infty}(\Omega)$ and there exists a constant C which doesn't depend on f, ϕ, u_1 such that

$$||\nabla u_1||_\infty \le C(|f|_p + ||\nabla\phi||_\infty). \tag{3.121}$$

Theorem 3.35. Let $\phi, \psi \in W^{1,\infty}(\Omega)$ satisfy (3.19), (3.20), $f \in L^p(\Omega)$ with $p > n$. If (3.17), (318) hold then under the above assumptions the solution u_2 of

$$u_2 \in K_\phi^\psi, \quad <-Au_2, v - u_2> \ge <f, v - u_2> \quad \forall v \in K_\phi^\psi$$

is in $W^{1,\infty}(\Omega)$ and there exists a constant C which doesn't depend on f, ϕ, ψ, u_2 such that

$$||\nabla u_2||_\infty \le C(|f|_p + ||\nabla\phi||_\infty + ||\nabla\psi||_\infty). \tag{3.122}$$

Proof: For β_1, β_2 satisfying (3.9), (3.10) and ϕ, ψ satisfying (3.19), (3.20) let us consider the solution u_ε of:

$$\begin{cases} -Au_\varepsilon + \dfrac{\beta_1}{\varepsilon}(u_\varepsilon - \phi) + \dfrac{\beta_2}{\varepsilon}(u_\varepsilon - \psi) = f \\[2mm] u_\varepsilon \in H_0^1(\Omega). \end{cases}$$

Introducing the solution w of

$$\begin{cases} -Aw = f \\[2mm] w \in H_0^1(\Omega) \end{cases}$$

we see that $u_\varepsilon - w$ satisfies (3.105) i.e.,:

$$\begin{cases} A(u_\varepsilon - w) = \dfrac{\beta_1}{\varepsilon}((u_\varepsilon - w) - (\phi - w)) + \dfrac{\beta_2}{\varepsilon}((u_\varepsilon - w) - (\psi - w)) \\[2mm] u_\varepsilon - w \in H_0^1(\Omega) \end{cases}$$

and by Proposition 3.32 we get

$$||\nabla(u_\varepsilon - w)||_\infty \leq C \text{ Max}(||\nabla(\phi - w)||_\infty, ||\nabla(\psi - w)||_\infty).$$

But by (3.18) and the Sobolev embedding theorems we have

$$||\nabla w||_\infty \leq C||u||_{2,p} \leq C|f|_p$$

and thus we get easily

$$||\nabla u_\varepsilon||_\infty \leq C(|f|_p + ||\nabla\phi||_\infty + ||\nabla\psi||_\infty), \tag{3.123}$$

where C doesn't depend on ε. Taking now $\beta_2 \equiv 0$, $\psi = |\phi|_\infty$, $\beta_1 < 0$ when $t < 0$ we get (3.121) from (3.123) by applying Theorem 3.4 and by taking the limit of (3.105) when $\varepsilon \to 0$. Similarly (3.122) and the Theorem 3.35 follow by choosing $\beta_1 < 0$ when $t < 0$, $\beta_2 > 0$ when $t > 0$ and by passing to the limit in (3.123).

<u>Remark 3.36.</u> It is clear from (3.121) that the mapping $\phi \to u_1(\phi) = u_1$ where u_1 is the solution of

$$u_1 \in K_\phi, \ <-Au_1, v-u_1> \geq 0 \ \ \forall v \in K_\phi = \{v \in H_0^1(\Omega)|v(x) \geq \phi(x) \tag{3.124}$$
$$\text{a.e. on } \Omega\}$$

is continuous at 0 on $W^{1,\infty}(\Omega)$ - i.e., if $\phi_n \in W^{1,\infty}(\Omega)$, $\phi_n \leq 0$ on Γ, $\phi_n \to 0$ in $W^{1,\infty}(\Omega)$ then $u_1(\phi_n) \to 0$ in $H_0^1(\Omega) \cap W^{1,\infty}(\Omega)$. But $\phi \to u_1(\phi)$ is not generally continuous in $W^{1,\infty}(\Omega)$ when $n > 1$. Indeed following [30] we easily see that in $\mathbb{R}^n$ ($n > 1$) and for

$$\Omega = \{x \in \mathbb{R}^n | \ |x| < 1\}, \quad A = \Delta, \quad \phi(x) = \varepsilon - r = \varepsilon - |x| \qquad (\varepsilon > 0)$$

the solution of (3.124) is given by (use similar arguments that those right after (3.45))

$$u_1(\phi) = \begin{cases} \varepsilon - r & \text{if } r \le 1 - \sqrt{1-\varepsilon} \\[2mm] (1 - \sqrt{1-\varepsilon})^2(\frac{1}{r} - 1) & \text{if } r \ge 1 - \sqrt{1-\varepsilon}. \end{cases}$$

Now clearly $||\nabla u_1(\phi)||_\infty = 1$. But when ε goes to 0, $\phi \to \phi_0(r) = -r$ in $W^{1,\infty}(\Omega)$ and the solution of (3.124) for ϕ_0 is given by $u_1(\phi_0) = 0$. Thus we cannot have $u_1(\phi) \to u_1(\phi_0)$.

A similar argument shows the impossibility of (3.120) in general. Indeed Ω being as above and A equal to Δ let us consider u_2 the solution of the two obstacle problem where $f = 0$ and the obstacles are given for instance by

$$\psi(x) = r = |x| \qquad \phi(x) = \varepsilon(\frac{1}{4} - |r - \frac{1}{2}|)^+.$$

(Then (3.120) implies that

$$||\nabla u_2||_\infty \le C \ \text{Max}(||\nabla \phi^+||_\infty, ||\nabla \psi^-||_\infty) = C'\varepsilon$$

and u_2 would go to 0 in $W^{1,\infty}(\Omega)$. This is impossible since one can show that u_2 touches ψ on a set of positive measure around 0.)

3.7. $W^{1,p}(\Omega)$-Regularity $(2 < p < +\infty)$.

The key arguments in this section will combine the results of 3.6 with interpolation techniques. So let us begin with a few words about interpolation. Our statements will be as simple as possible, and we will just explain what is useful for our purpose. Details and extensions can be found in [20], [39] or [108].

Let A_0, A_1 be two Banach spaces with norms $|\ |_{A_0}$ and $|\ |_{A_1}$ respectively and such that

$$A_1 \subset A_0.$$

We will confine ourself to the K-interpolation method. So for $t \in \mathbb{R}^+$, $a \in A_0$ let us define K by

$$K(t,a) = \inf_{\substack{a_i \in A_i \\ a_0 + a_1 = a}} (|a_0|_{A_0} + t|a_1|_{A_1}).$$

Then for $0 < \theta < 1$, $1 \leq q < +\infty$ (other values of θ,q are suitable but we will not use them here) set

$$(A_0,A_1)_{\theta,q} = \{a \in A_0 | t^{-\theta}K(t,a) \in L^q(\mathbb{R}^+, \tfrac{dt}{t})\}.$$

($L^q(\mathbb{R}^+, \tfrac{dt}{t})$ denotes the space of equivalence classes of functions of the qth power integrable with respect to the measure dt/t.)

For $a \in (A_0,A_1)_{\theta,q}$ we will set

$$|a|_{\theta,q} = \left[\int_{\mathbb{R}^+} [t^{-\theta}K(t,a)]^q \frac{dt}{t}\right]^{1/q}. \tag{3.125}$$

Then we first can prove that $(A_0,A_1)_{\theta,q}$ gives a family of "interpolation" spaces between A_0 and A_1, that is to say we have:

__Theorem 3.37.__ Under the above assumptions $(A_0,A_1)_{\theta,q}$ is a Banach space for the norm $| \ |_{\theta,q}$ and we have the algebraical and topological embeddings:

$$A_1 \subset (A_0,A_1)_{\theta,q} \subset A_0.$$

__Proof:__ See [20], [39] or [108].

Now it is well known that linear bounded operators mapping a pair of Banach spaces to another pair also map continuously their interpolation spaces. As we will see, this result extends to the case of Lipschitz functions. (For more general results, see [104].)

So let A_1,A_0,B_1,B_0 Banach spaces with norms denoted by $| \ |_{A_i}$, $| \ |_{B_i}$ such that

$$A_1 \subset A_0, \quad B_1 \subset B_0.$$

Then we have:

__Theorem 3.38.__ Under the above assumptions, let T be a map from A_0 to B_0 such that $T(A_1) \subset B_1$. Let us assume that there exist two constants C and C' such that

$$|T(a_0) - T(a_0')|_{B_0} \leq C|a_0 - a_0'|_{A_0} \quad \forall a_0, a_0' \in A_0$$

$$|T(a_1)|_{B_1} \leq C'|a_1|_{A_1} \quad \forall a_1 \in A_1.$$

Then for all $a \in (A_0,A_1)_{\theta,q}$, $T(a) \in (B_0,B_1)_{\theta,q}$ and we have

$$|T(a)|_{\theta,q} \leq \text{Max}(C,C')|a|_{\theta,q}.$$

Proof: Let us write

$$T(a) = (T(a) - T(a_1)) + T(a_1) \quad \forall a_1 \in A_1, \ \forall a \in A_0.$$

By definition of K (between B_0 and B_1) we get

$$K(t,T(a)) \leq |T(a) - T(a_1)|_{B_0} + t|T(a_1)|_{B_1}$$

$$\leq C|a - a_1|_{A_0} + tC'|a_1|_{A_1}$$

$$\leq \text{Max}(C,C')[|a - a_1|_{A_0} + t|a_1|_{A_1}].$$

Since this inequality holds for all $a_1 \in A_1$ taking the Inf on a_1 we obtain:

$$K(t,T(a)) \leq \text{Max}(C,C')K(t,a)$$

and the result follows from our definition of the interpolation spaces and from (3.125). Note that we have used the same notation for the norm $| \ |_{\theta,q}$ on $(A_0,A_1)_{\theta,q}$ and $(B_0,B_1)_{\theta,q}$ but there should be no confusion.

In the framework of the one obstacle problem we now want to apply the above results with

$$A_0 = B_0 = W_0^{1,2}(\Omega), \quad A_1 = B_1 = W_0^{1,\infty}(\Omega)$$

where we have set (see (2.4)) $W_0^{1,\infty}(\Omega) = W_0^{1,2}(\Omega) \cap W^{1,\infty}(\Omega)$ with the norm $\| \ \|_{1,\infty}$.

As a consequence of the interpolation results of [56] we have if Ω is a bounded, Lipschitz open set of $\mathbb{R}^n$:

Theorem 3.39. For all p such that $2 < p < +\infty$

$$(W_0^{1,2}(\Omega), W_0^{1,\infty}(\Omega))_{1-\frac{2}{p},p} = W_0^{1,p}(\Omega).$$

Proof. See [56].

From now on, we will assume that Ω is a smooth, bounded open set of $\mathbb{R}^n$ (for instance of class C^2) and that the operator A defined by (3.2) satisfies (3.1) and has its coefficients in $C^1(\bar{\Omega})$.

Thus we have (see [23]).

Theorem 3.40. Let $f \in W^{-1,p}(\Omega)$, $2 < p < +\infty$. Then for all $\phi \in W^{1,p}(\Omega)$, $\phi \leq 0$ on Γ, the solution u_1 of

$$u_1 \in K_\phi, \quad <-Au_1,v - u_1> \geq <f,v - u_1> \quad \forall v \in K_\phi \qquad (3.126)$$

is in $W_0^{1,p}(\Omega)$ and we have for some constant C which doesn't depend on
ϕ, f, u_1

$$||u_1||_{1,p} \leq C(||f||_{-1,p} + ||\phi||_{1,p}) \qquad (3.127)$$

$||f||_{-1,p}$ denoting the strong dual norm on $W^{-1,p}(\Omega)$.

<u>Proof</u>: Let us assume first that $f = 0$. Then (see (3.7)) one can replace
ϕ by ϕ^+ so that in this case one can assume $\phi \in W_0^{1,p}(\Omega)$. Thus for
$\phi \in W_0^{1,2}(\Omega)$ let us consider

   T: $\phi \to T(\phi)$,

where $T(\phi)$ is the solution of (3.126) with $f = 0$.

   For $\phi, \phi' \in H_0^1(\Omega)$ we have $T(\phi) \geq \phi$ and $T(\phi') \geq \phi'$ thus

$$T(\phi) + (\phi' - \phi) \geq \phi', \quad T(\phi') + (\phi - \phi') \geq \phi,$$

and the above functions are all in $H_0^1(\Omega)$. Thus using the inequalities

   $<-A(T(\phi)), v - T(\phi)> \geq 0 \quad \forall v \in K_\phi$

   $<-A(T(\phi')), v - T(\phi')> \geq 0 \quad \forall v \in K_{\phi'}$

with v equal to $T(\phi') + (\phi - \phi')$ in the first and $v = T(\phi) + (\phi' - \phi)$
in the second one leads to

   $<-A(T(\phi)),(T(\phi') - T(\phi)) + (\phi - \phi')> \geq 0$

   $<-A(T(\phi')),(T(\phi) - T(\phi')) + (\phi' - \phi)> \geq 0.$

By adding these we get

   $<-A(T(\phi) - T(\phi')),T(\phi) - T(\phi')> \leq <-A(T(\phi) - T(\phi')),(\phi - \phi')>$

and (3.1) with the Cauchy-Schwarz inequality leads to

$$||T(\phi) - T(\phi')||_{1,2} \leq C||\phi - \phi'||_{1,2}. \qquad (3.128)$$

Now if $\phi \in W_0^{1,\infty}(\Omega)$ by Theorem 3.34 we have $T(\phi) \in W_0^{1,\infty}(\Omega)$ and

$$||T(\phi)||_{1,\infty} \leq C'||\phi||_{1,\infty}.$$

Thus these two last inequalities combined with Theorems 3.38 and 3.39 give
us that for $\phi \in W_0^{1,p}(\Omega)$, $T(\phi) \in W_0^{1,p}(\Omega)$ and an estimate

$$||T(\phi)||_{1,p} \leq C||\phi||_{1,p}.$$

This proves the theorem when $f = 0$. For the general case introduce the solution w of

$$\begin{cases} -Aw = f \\ w \in H_0^1(\Omega). \end{cases}$$

We can check (see the proof of Theorem 3.15) that $u_1 - w$ is the solution of (3.126) with $\phi - w$ in place of ϕ. But from our study of the case where $f = 0$, and since by regularity result for the Dirichlet problem $w \in W_0^{1,P}(\Omega)$, it results that $u_1 - w \in W_0^{1,P}(\Omega)$ with the estimate

$$||u_1 - w||_{1,p} \leq C||\phi - w||_{1,p},$$

and (3.127) follows from the fact that we have

$$||w||_{1,p} \leq C||f||_{-1,p}.$$

<u>Remark 3.41</u>. In the case of two obstacles, due to difficulty of obtaining the equivalent of (3.128) and due to the fact that in this case it is not possible to assume that $\phi = \psi = 0$ on Γ, the above technique doesn't apply.

To conclude this section, let us take a look at the continuity in $W^{1,P}(\Omega)$ of the mapping $\phi \to u_1(\phi)$ where $u_1(\phi) = u_1$ is the solution of (3.126). For f fixed in $W^{-1,P}(\Omega)$ the problem of the continuity of $\phi \to u_1(\phi)$ from $W^{1,P}(\Omega)$-strong into $W^{1,P}(\Omega)$-strong $(2 < p < +\infty)$ is actually open (see Remark 3.44 below). However, some interesting results have been obtained by L. Boccardo - F. Murat [24]. In particular, one can prove:

<u>Theorem 3.42</u>. Let f be fixed in $H^{-1}(\Omega)$ and ϕ_n be a sequence of obstacles such that $\phi_n \in H_0^1(\Omega)$. If $\phi_n \longrightarrow \phi$ in $W^{1,P}(\Omega)$ weak $(2 < p < +\infty)$ when $n \to +\infty$, then

$$u_1(\phi_n) \to u_1(\phi) \quad \text{in} \quad H_0^1(\Omega) \quad \text{strong}.$$

To prove Theorem 3.42 we will use the following result (see [93], [31] for a proof).

<u>Theorem 3.43</u>. Let Ω be a bounded open set in $\mathbb{R}^n$. If $f_n \in H^{-1}(\Omega)$ is a sequence such that $f_n \geq 0$ (in the measure or distributional sense), $f_n \longrightarrow f$ in $H^{-1}(\Omega)$-weak, then $f_n \to f$ in $W^{-1,q}(\Omega)$ strong for all $q < 2$.

<u>Proof of Theorem 3.42</u>: Let $u_1^n = u_1(\phi^n)$ be the solution of

$$\begin{cases} u_1^n \in K_{\phi_n} = \{v \in H_0^1(\Omega)\,|\,v(x) \geq \phi_n(x) \quad \text{a.e. in } \Omega\} \\[2mm] <-Au_1^n, v - u_1^n> \,\geq\, <f, v - u_1^n> \quad \forall v \in K_{\phi_n}. \end{cases} \tag{3.129}$$

Taking $v = \phi_n$ in (3.129) we easily obtain

$$||u_1^n||_{1,2} \leq C$$

where C doesn't depend on n. (We note that by our assumptions $||\phi_n||_{1,2}$ is bounded independently of n.) Thus, (see Theorem 2.5), there exists a subsequence of (u_1^n, ϕ_n) still denoted by (u_1^n, ϕ_n) such that

$$u_1^n \rightharpoonup u \quad \text{in } H_0^1(\Omega)$$

$$u_1^n \to u \quad \text{in } L^2(\Omega) \quad \text{and pointwise a.e.}$$

$$\phi^n \to \phi \quad \text{in } L^2(\Omega) \quad \text{and pointwise a.e.}$$

From $u_1^n(x) \geq \phi_n(x)$ a.e. we easily deduce that $u \in K_\phi$. Moreover, re-writing (3.129) as

$$<-Au_1^n - f, v - u_1^n> \,\geq\, 0 \quad \forall v \in K_{\phi_n}$$

and taking $v = u_1^n + \xi$ where $\xi \geq 0$, $\xi \in \mathscr{D}(\Omega)$ we obtain that

$$-Au_1^n - f \geq 0 \quad \text{in } \Omega.$$

Applying Theorem 3.43 we obtain (using for instance the fact that a limit is unique in $\mathscr{D}'(\Omega)$)

$$-Au_1^n - f \to -Au - f \quad \text{in } W^{-1,p'}(\Omega)\,(\tfrac{1}{p} + \tfrac{1}{p'} = 1). \tag{3.130}$$

Now taking for $w \in K_\phi$, $v = w + \phi_n - \phi$ in (3.129) we have

$$<-Au_1^n, w> \,\geq\, <-Au_1^n, u_1^n> + <f, w - u_1^n> - <-Au_1^n - f, \phi_n - \phi>. \tag{3.131}$$

Taking the limit - inf of both sides we obtain using the lower semicontinuity of $u \to <-Au, u>$ and the fact that $<-Au_1^n - f, \phi_n - \phi> \to 0$ (this is due to (3.130) and from our assumption on ϕ_n)

$$<-Au, w> \,\geq\, <-Au, u> + <f, w - u> \quad \forall w \in K_\phi.$$

This proves that $u = u_1$ and the weak convergence of u_1^n toward u_1. The strong convergence is obtained by noting first that:

$$\nu||u_1^n-u_1||_{1,2}^2 \le <-A(u_1^n-u_1),u_1^n - u_1> = <-Au_1^n,u_1^n-u_1> - <-Au_1,u_1^n-u_1>.$$

Then applying (3.131) with $w = u_1$ we deduce:

$$\nu||u^n-u_1||_{1,2}^2 \le <f,u_1^n-u_1> + <-Au_1^n-f,\phi_n - \phi> - <-Au_1,u_1^n-u_1> \to 0$$

with n and the result follows.

Remark 3.44. As a consequence of Theorem 3.42 and under the same assumptions we obtain also (see [24])

$$u_1(\phi_n) \to u_1(\phi) \quad \text{in} \quad W_0^{1,q}(\Omega) \quad \text{strong}$$

for all q satisfying $2 < q < p$.

Indeed, as a consequence of Hölder's inequality, we have

$$||\nabla(u_1(\phi_n) - u_1(\phi))||_q \le ||\nabla(u_1(\phi_n) - u_1(\phi))||_2^r ||\nabla(u_1(\phi_n) - u_1(\phi))||_p^s$$

$$\le C||\nabla(u_1(\phi_n) - u_1(\phi))||_2^r \quad \text{(by (3.127))}$$

with $r = \dfrac{2(p-q)}{q(p-2)}$ and $s = \dfrac{p(q-2)}{q(p-2)}$.

Appendix

Lᵖ–Estimates for the Solution of the Dirichlet Problem

In this section $\mathcal{O}$ will be a bounded open set of $\mathbb{R}^n$ and a_{ij} will be functions in $L^\infty(\mathcal{O})$ satisfying the usual coerciveness assumption

$$a_{ij}(x) \cdot \xi_i \xi_j \geq \nu |\xi|^2 \qquad \forall x \in \mathcal{O} , \quad \forall \xi \in \mathbb{R}^n \qquad (A.1)$$

with $\nu > 0$. A will be the operator defined by (3.2).

Theorem A_1. (Stampacchia [100]). Under the above assumptions, if $f_0, \ldots, f_n$ are functions in $L^p(\mathcal{O})$ with $p \geq 2$ and u the solution of the Dirichlet problem

$$\begin{cases} -Au = f_0 + \sum_{i=1}^{n} \dfrac{\partial f_i}{\partial x_i} & \text{in } \mathcal{O} \\[2mm] u \in H_0^1(\mathcal{O}). \end{cases} \qquad (A.2)$$

Then

(i) If $p > n$, we have $u \in L^\infty(\mathcal{O})$ and u satisfies an estimate of type

$$|u|_\infty \leq C \sum_{i=0}^{n} |f_i|_p.$$

(ii) If $2 \leq p < n$, then we have $u \in L^{p^*}(\mathcal{O})$ where p^* is defined by $\dfrac{1}{p^*} = \dfrac{1}{p} - \dfrac{1}{n}$ and u satisfies an estimate of type

$$|u|_{p^*} \leq C \sum_{i=0}^{n} |f_i|_p.$$

(iii) If $p = n = 2$, then we have $u \in L^q(\mathcal{O})$ for all q and

$$|u|_q \leq C \sum_{i=0}^{n} |f_i|_2.$$

The constants C in (i), (ii), and (iii) are independent of u and of the f_i; $|\ |_p$ denotes L^p-norms in $L^p(\mathscr{O})$.

Proof: First, let us assume that $p = 2$. Then from (A.1), (A.2), and (2.7), we deduce that

$$\nu||\nabla u||_2^2 \leq \int_{\mathscr{O}} a_{ij} u_{x_j} u_{x_i} = \int_{\mathscr{O}} f_0 u - f_i u_{x_i} \leq C \sum_{i=0}^{n} |f_i|_2 \cdot ||\nabla u||_2. \qquad (A.3)$$

Hence, it follows that

$$||\nabla u||_2 \leq \frac{C}{\nu} \cdot \sum_{i=0}^{n} |f_i|_2. \qquad (A.4)$$

Thus, if $p = 2 = n$, (iii) follows immediately from Sobolev's inequality (see for instance [82]). Now if $n = 1$, $\mathscr{O} = (x_0, x_1)$, we have

$$|u(x)| \leq \int_{x_0}^{x} |u'(t)| dt \leq |u'|_2 |\Omega|^{1/2}$$

and this combined with (A.4) gives

$$|u|_\infty \leq C \sum_{i=0}^{n} |f_i|_2 \leq C' \sum_{i=0}^{n} |f_i|_p$$

which proves (i) in this case. Thus we now assume $n \geq 2$ and that we are not in the case (iii).

Let us consider, for $k > 0$, ξ defined by

$$\xi = (u - k)^+ - (u + k)^-.$$

As the sum of two functions of $H_0^1(\mathscr{O})$, it is clear (see Theorem 2.4) that ξ is a function in $H_0^1(\mathscr{O})$ which vanishes on $|u| \leq k$ and is equal to $\pm(|u| - k)$ on $|u| > k$. Let us denote by $A(k)$ the set

$$A(k) = \{x \in \mathscr{O} \mid |u(x)| > k\}.$$

An application of the same arguments as in (A.3) with Hölder's inequality leads to

$$\nu||\nabla \xi||_2^2 \leq \int_{\mathscr{O}} a_{ij} \xi_{x_j} \xi_{x_i} = \langle -Au, \xi \rangle = \int_{\mathscr{O}} f_0 \xi$$

$$- f_i \xi_{x_i} \leq C \sum_{i=0}^{n} |f_i|_p \cdot ||\nabla \xi||_{p'} \qquad (A.5)$$

where C depends only on $\mathcal{O}$. But now since $p \geq 2$, that is to say, $p' \leq 2$, we have by Hölder's inequality

$$||\nabla\xi||_{p'}^{p'} = \int_{A(k)} |\nabla\xi|^{p'} \leq \left(\int_{A(k)} |\nabla\xi|^{p' \cdot 2/p'}\right)^{p'/2} |A(k)|^{1-(p'/2)}$$

$$\iff ||\nabla\xi||_{p'} \leq ||\nabla\xi||_2 |A(k)|^{\frac{1}{p'} - \frac{1}{2}}$$

where $|A(k)|$ denotes the measure of $A(k)$. From (A.5), we then deduce that

$$||\nabla\xi||_{p'} \leq C \sum_{i=0}^{n} |f_i|_p \cdot |A(k)|^{\frac{2}{p'} -1}$$

But using now the Sobolev's inequality (2.8), we deduce that

$$|\xi|_{p'*} \leq C \sum_{i=0}^{n} |f_i|_p \cdot |A(k)|^{\frac{2}{p'} -1} \qquad (A.6)$$

with $\frac{1}{p'*} = \frac{1}{p'} - \frac{1}{n} = 1 - \frac{1}{p} - \frac{1}{n}$ and C depending only on $\mathcal{O}$ and p. (Let us remark that $p'*$ is well defined. Indeed, if we have $p' = n$, since $n \geq 2$, we would have $p' = p = 2 = n$ which is the case (iii).) Now, let us note that for $h > k$, we have $A(h) \subset A(k)$ and so for such a h, we have

$$(h-k)|A(h)|^{1/p'*} \leq \left[\int_{A(h)} (|u| - k)^{p'*}\right]^{1/p'*}$$

$$\leq \left[\int_{A(k)} (|u| - k)^{p'*}\right]^{1/p'*} = |\xi|_{p'*} \leq C \sum_{i=0}^{n} |f_i|_p |A(k)|^{\frac{2}{p'} -1}.$$

This can be written as

$$|A(h)| \leq \left(C \sum_{i=0}^{n} |f_i|_p / (h-k)\right)^{p'*} |A(k)|^{p'*(\frac{2}{p'} -1)}. \qquad (A.7)$$

Let us now use the following lemma which will be proved later on.

<u>Lemma.</u> Let $\phi: [0,+\infty) \to \mathbb{R}^+$ be a nonincreasing function satisfying

$$\phi(h) \leq \left(\frac{c}{h-k}\right)^{\alpha} \phi(k)^{\beta} \qquad h > k \qquad (A.8)$$

where c, α, β are positive constants. Then

(i) If $\beta > 1$, we have $\phi(d) = 0$ for $d = c \cdot \phi(0)^{\beta-1/\alpha} \cdot 2^{\beta/\beta-1}$

(ii) if $\beta < 1$, we have $\phi(h) \leq \left(\dfrac{2^{1/1-\beta} c}{h}\right)^{\mu}$ with $\mu = \alpha/1 - \beta$ for

$h > 0$.

More precisely, if we put $\phi(t) = |A(t)|$, then we have by (A.7) an inequality of type (A.8). Now if $p > n$, the exponent $\beta = p'* \cdot (\frac{2}{p'} - 1) =$
$(1 - \frac{2}{p})/(1 - \frac{1}{p} - \frac{1}{n})$ is greater than 1 and by (i) of the lemma we deduce
that $\phi(d) = 0$ (i.e. $|u| \leq d$ a.e.) with

$$d = C \sum_{i=0}^{n} |f_i|_p \cdot |\mathcal{O}|^{\frac{\beta-1}{p'*}} \cdot 2^{\frac{\beta}{\beta-1}} .$$

Hence (i) results. Now if $p < n$, we have $\beta = p'* (\frac{2}{p'} - 1) < 1$ and by the
part (ii) of the lemma and since

$$\mu = \frac{p'*}{1 - p'*(\frac{2}{p'} - 1)} = \frac{1}{\frac{1}{p'*} - \frac{2}{p'} + 1} = \frac{1}{\frac{1}{p'} - \frac{1}{n} - \frac{2}{p'} + 1} = \frac{1}{\frac{1}{p} - \frac{1}{n}} = p* ,$$

we deduce that

$$|[|u| > h]| \leq \left(\frac{C \cdot \sum_{i=0}^{n} |f_i|_p}{h}\right)^{p*} . \tag{A.9}$$

Consider now for $i = 0,\ldots,n$ the mapping $T_i : f \to u_i = T_i(f)$ where u_i
is the solution of

$$-Au_i = \frac{\partial f}{\partial x_i}$$

$$u_i \in H_0^1(\mathcal{O})$$

(with $\frac{\partial f}{\partial x_0} = f$). Then by (A.9) for all $2 \leq q < n$, T_i is a linear map
from $L^q(\mathcal{O})$ into $L^{q*}(\mathcal{O})$-weak. Choose, for instance, $2 \leq p < q < n$,
then T_i maps $L^2(\mathcal{O})$ into $L^{2*}(\mathcal{O})$ and $L^q(\mathcal{O})$ into $L^{q*}(\mathcal{O})$-weak.
Therefore by the Marcinkiewicz Theorem (see [20], p. 6) T_i maps $L^p(\mathcal{O})$
into $L^{p*}(\mathcal{O})$ continuously and we have by (A.9)

$$|T_i(f)|_{p*} \leq C|f|_p .$$

Since now $u = \sum_{i=0}^{n} T_i(f_i)$, (ii) follows. What remains now is to prove
the lemma.

<u>Proof of the Lemma</u>: By (A.8), we have for $h_n = d - \dfrac{d}{2^n}$ $(n \in \mathbb{N})$

$$\phi(h_{n+1}) \leq \left(\frac{c}{h_{n+1} - h_n}\right)^{\alpha} \phi(h_n)^{\beta} = \frac{c^{\alpha}}{d^{\alpha}} 2^{(n+1)\alpha} \phi(h_n)^{\beta} .$$

Now we prove by induction that $\phi(h_n) \leq \phi(0)2^{\frac{n\alpha}{1-\beta}}$. If $n = 0$, it is clear. If it is true at order n, from the inequality above we deduce

$$\phi(h_{n+1}) \leq \frac{c^\alpha}{d^\alpha} \cdot 2^{(n+1)\alpha}\phi(0)^\beta \cdot 2^{\frac{n\alpha\beta}{1-\beta}}.$$

But by definition of d this leads to

$$\phi(h_{n+1}) \leq \phi(0)2^{(n+1)\alpha} \cdot 2^{\frac{n\beta\alpha}{1-\beta}} \cdot 2^{\frac{-\alpha\beta}{\beta-1}} = \phi(0)2^{\frac{(n+1)\alpha}{1-\beta}}.$$

Thus, we have

$$0 \leq \phi(d) \leq \phi(h_n) \leq \phi(0)\, 2^{\frac{n\alpha}{1-\beta}} \to 0 \quad \text{when} \quad n \to +\infty \quad \text{if} \quad \beta > 1$$

and (i) results. For (ii) put $\psi(h) = (\frac{h}{c})^\mu \phi(h)$ with $\mu = \frac{\alpha}{1-\beta} > 0$. By (A.8), since $\phi(h) = (\frac{c}{h})^\mu \psi(h)$, we have for $h > k > 0$

$$(\frac{c}{h})^\mu \psi(h) \leq (\frac{c}{h-k})^\alpha (\frac{c}{k})^{\mu\beta} (\psi(k))^\beta$$

$$\text{(A.10)}$$

$$\Longleftrightarrow \psi(h) \leq \frac{h^\mu}{(h-k)^\alpha k^{\mu\beta}}[\psi(k)]^\beta.$$

Moreover, since ϕ is bounded by $\phi(0), \psi(h) \to 0$ when $h \to 0$. So let T be such that

$$0 \leq \psi(t) \leq 1 \qquad \forall t \in (0,T).$$

Let h be in $(0,T)$, from (A.10), we deduce that for all n

$$\psi(2^n h) \leq \frac{(2^n h)^\mu}{(2^n h - 2^{n-1} h)^\alpha (2^{n-1} h)^{\mu\beta}}[\psi(2^{n-1} h)]^\beta$$

$$\Longleftrightarrow \psi(2^n h) \leq 2^\mu [\psi(2^{n-1} h)]^\beta.$$

And by iteration this leads to

$$\psi(2^n h) \leq 2^{\mu(1+\beta+\ldots+\beta^{n-1})}[\psi(h)]^{\beta^n}.$$

Since $h \in (0,T)$, we have

$$\psi(2^n h) \leq 2^{\frac{\mu}{1-\beta}} \qquad \forall n \in \mathbb{N} \qquad \forall h \in (0,T)$$

and so ψ is bounded on $(0,+\infty)$ by $2^{\frac{\mu}{1-\beta}}$ which gives the result.

Remark 1. Let us note that the theorem can be slightly improved by taking $f_0 \in L^q(\mathcal{O})$, $\frac{1}{q} = \frac{1}{p} + \frac{1}{n}$ and by replacing $|f_0|_p$ by $|f_0|_q$ in all inequalities (indeed it suffices in (A.5) to apply Hölder's inequality and the Sobolev embedding theorem).

As a corollary we have:

Corollary A_2. Under the assumptions of the Theorem A_1 and if u satisfies

$$-Au \leq f_0 + \sum_{i=1}^{n} \frac{\partial f_i}{\partial x_i} \quad \text{in } \mathcal{O} \quad \text{(in a weak sense)}$$

$$u \geq 0 \quad \text{in } \mathcal{O} \tag{A.10}$$

$$u \in H_0^1(\mathcal{O})$$

then (i), (ii), and (iii) hold without any change.

Proof: Let us call v the solution of (A.2). Then from (A.10), we have

$$-Au \leq -Av. \tag{A.11}$$

By considering $(u - v)^+$, we have from (A.11)

$$\nu ||\nabla(u-v)^+||_2^2 \leq <-A(u-v), (u-v)^+> \leq 0$$

and so the result follows from the Theorem A_1 since we have

$$0 \leq u \leq v.$$

Remark 2. Note that the above results are true without assumptions of regularity on $\mathcal{O}$ and a_{ij} except that they have to be bounded.

COMMENTS

Before commenting on the regularity theory itself, let us note that an axiomatisation of the penalty method can be found in Lions [83].

Regularity in $W^{2,p}(\Omega)$ ($2 < p < +\infty$) has been initiated for obstacle problems by Lewy-Stampacchia [80] (c.f. also, [76] for a description of this method). Other proofs and results are available in Brezis-Stampacchia [37] and Stampacchia [101] and also via the Lewy-Stampacchia inequality (see [81]) in Mosco - Troianiello [92].

For $W^{2,\infty}(\Omega)$-regularity the pioneering papers are those of Brezis-Kinderlehrer [34] and Gehrardt [66]. Caffarelli-Kinderlehrer [46] give other techniques using potential theory.

Let us note that $W^{2,\infty}(\Omega)$-regularity is a useful tool to study the free boundary of the problem. We refer the reader to Caffarelli [40] and Kinderlehrer et. al. [74], [75], and [76] for this kind of result.

In the case of variable coefficients $W^{1,\infty}(\Omega)$-regularity theory follows Chipot [50], see also [54] for other variants. Other techniques and ideas for $C^{0,\alpha}(\Omega)$-regularity can be found in [22], [46] and [64], [96] when $\alpha = 1$.

An $L^p(\Omega)$-regularity theory is done in Adams [2]. See also Adams [3] for interesting results via capacity theory. Finally we note that we have not included here regularity theory for thin obstacle problems. The reader may refer mainly to [41], [60], [70], and [73].

Chapter 4
The Dam Problem

Among the applications of Variational Inequalities with obstacles, the most famous one is probably that of the Dam Problem. The modern approach started with the pioneering work of C. Baiocchi (see [9], [10]). Many developments have followed, many of them due to the Pavia school, (see [11], [12], [16], [17] and [13], [15] where a complete bibliography can be found). Unfortunately this approach is only possible in the case of porous media with vertical walls. For instance, the case of a rectangular dam (as in the figure (4.1(B)) is suitable in dimension 2 (see [76], [102] for the case of porous media in three dimensions). To overcome this restriction several attempts have been made including formulations via Quasivariational Inequalities (see [11], [12], [15]) and the use of a "maximal" solution (see [5]) (see also [109] for a very interesting approach). More recently H. Brezis, D. Kinderlehrer, and G. Stampacchia in [35] and H. W. Alt in [7] introduced a new formulation which can treat the case of general domains. We shall adopt the setting of [35] and we will show how it can be reduced to a Variational Inequality with obstacle in the rectangular case. Let us first begin by presenting the physical problem (see [32], [49]).

4.1. Statement of the Problem

4.1.1. Notation

We will restrict ourselves to the case of two dimensions; Ω will be a bounded domain with a locally Lipschitz continuous boundary S. Ω represents the section of a porous medium.

We subdivide S into the following subsets (see Fig. (4.1)):

74

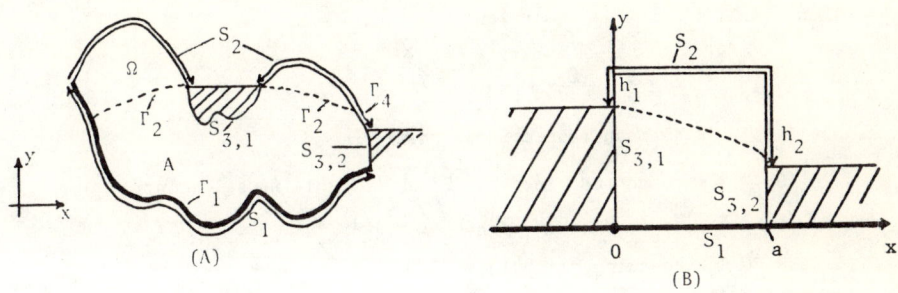

Figure 4.1

S_1 the impervious part of the dam

S_2 the part of the boundary $\partial\Omega$ of Ω in contact with the open air

S_3 the part of the dam under water.

For convenience in some theorems we shall assume that S_1 is a closed subset of $\partial\Omega$ and that S_3 is open in the complement of S_1 in S. Moreover, in the case of several reservoirs we shall also denote by $S_{3,1}, S_{3,2}, \ldots, S_{3,n}$ the different connected components of S_3. For instance, in the above figure $S_3 = S_{3,1} \cup S_{3,2}$.

Now, instead of giving directly the proof of our theorem in full generalities we shall often restrict ourselves to the following simpler case:

The domain Ω will be assumed to be vertically convex, that is to say Ω will satisfy:

$$\forall (x,y),(x,y') \in \Omega \text{ the segment } \{x\} \times [y,y'] \subset \Omega. \qquad (4.2)$$

Moreover we will assume that:

S_1 is a closed connected subset of S which surrounds
Ω from below   (4.3)

$S_2 \cup S_3$ surrounds Ω from above   (4.4)

S has only a finite number of vertical walls.   (4.5)

Note that both examples given in the Figure (4.1) satisfy these assumptions.

More precisely, if we denote by π_x the usual projection of $\mathbb{R}^2$ on the x axis, for $x \in \pi_x(\Omega)$, we can define

$$S^-(x) = \text{Inf}\{y \mid (x,y) \in \Omega\}, \quad S^+(x) = \text{Sup}\{y \mid (x,y) \in \Omega\}. \tag{4.6}$$

Then, (4.2) is clearly equivalent to

$$\Omega = \{(x,y) \mid x \in \pi_x(\Omega), \ S^-(x) < y < S^+(x)\}. \tag{4.7}$$

Now (4.3), (4.4) mean respectively that

S_1 is a connected piece of S which contains all points
(x,y) of S such that $y \le S^-(x)$ (4.8)

$S_2 \cup S_3$ contains all points (x,y) of S such that
$y \ge S^+(x)$. (4.9)

Moreover, (4.5) implies clearly that

S^- is continuous on $\pi_x(\Omega)$ except perhaps on a finite
subset $\mathscr{S}^-$ (4.10)

S^+ is continuous on $\pi_x(\Omega)$ except perhaps on a finite
subset $\mathscr{S}^+$. (4.11)

(Note that vertical walls can occur on $\mathscr{S}^- \cup \mathscr{S}^+$ and eventually at
the end of the dam; of course at these points S^-, S^+ have a right and
left limit -- when it makes sense.)

These assumptions are by no means necessary to study the mathematical problem in question. However, they should help the reader in a first reading. After each statement under the assumptions (4.2)-(4.5) we will indicate how to treat the general case.

The basic problem now is to find the pressure $p = p(x,y)$ of the fluid in Ω and that portion of Ω which is wet -- i.e., the wet set A.

4.1.2. Strong Formulation

The boundary of A is divided in four parts (see the Figure (4.1(A)). First Γ_1 the impervious part, then Γ_2 the free boundary of A, $\Gamma_3 = S_3$ the part covered by the fluid and finally Γ_4 the wet part of the dam in contact with the air. Among these pieces of the boundary, only Γ_3 is completely known.

Experimentally, the velocity $\vec{v}$ of the fluid, which we assume for instance to be water with a specific weight equal to 1, is given in A by Darcy's law

$$\vec{v} = -k\nabla(p + y)$$

with k being the coefficient of permeability and ∇ the operator $(\frac{\partial}{\partial x}, \frac{\partial}{\partial y})$.

If the medium is assumed to be homogeneous, then k is a constant which we shall choose strictly positive and in this case the incompressibility of the fluid leads to

$$\text{div}(\vec{v}) = 0 \quad \text{in A.}$$

Hence

$$\Delta p = 0 \quad \text{in A.} \tag{4.12}$$

Assume now that the atmospheric pressure is chosen to be 0 and neglect the effects of capillarity and of evaporation. Then if we denote by h_i (i = 1,...,n) the level of the water in the reservoir with bottom $S_{3,i}$ the pressure on $S_2 \cup S_3$ has to be given by the Lipschitz continuous function ϕ, where

$$\phi(x,y) = \begin{cases} 0 & \text{on } S_2 \\[2ex] h_i - y & \text{on } S_{3,i} \quad \forall i = 1,\ldots,n. \end{cases} \tag{4.13}$$

Thus p must fulfill the following Dirichlet boundary conditions

$$p = 0 \quad \text{on } \Gamma_2 \cup \Gamma_4, \quad p = \phi \quad \text{on } \Gamma_3. \tag{4.14}$$

Moreover the fact that no water can flow through Γ_1 and Γ_2 leads to conditions of Neumann type viz:

$$\vec{v} \cdot \vec{v} = -k\vec{v}(p + y) \cdot \vec{v} = 0 \quad \text{on } \Gamma_1 \cup \Gamma_2,$$

where $\vec{v} = (v_x, v_y)$ denotes the unit outward normal to A. This also gives:

$$\frac{\partial}{\partial v}(p + y) = 0 \quad \text{on } \Gamma_1 \cup \Gamma_2. \tag{4.15}$$

Finally, we can express the fluid flow through Γ_4 by

$$\vec{v} \cdot \vec{v} \geq 0 \quad \text{on } \Gamma_4,$$

and thus, since k is strictly positive, by

$$\frac{\partial}{\partial v}(p + y) \leq 0 \quad \text{on } \Gamma_4. \tag{4.16}$$

The problem is now to find (p,A) such that (4.12), (4.14), (4.15), and (4.16) hold.

4.1.3. Weak Formulation

We introduce here the ideas of H. Brezis, D. Kinderlehrer, and G. Stampacchia [35] (compare with [7]). Assume that we have found a pair (p,A) which is a solution of (4.12), (4.14), (4.15) and (4.16) with p and the boundary of A smooth enough. By applying the Green's formula, we have for $\xi \in C^1(\overline{\Omega})$ (see Section 2.1):

$$\int_A \nabla p \cdot \nabla \xi + \xi_y = \int_A -\Delta p \cdot \xi + \int_{\partial A} \frac{\partial}{\partial \nu}(p + y) \cdot \xi.$$

(We have denoted by ξ_y the derivative in the y direction. Here and in the following we omit the measures in the integrals). From (4.12), (4.15), we obtain

$$\int_A \nabla p \cdot \nabla \xi + \xi_y = \int_{\Gamma_3 \cup \Gamma_4} \frac{\partial}{\partial \nu}(p + y) \cdot \xi.$$

If we choose now $\xi = 0$ on Γ_3 and $\xi \geq 0$ on S_2 ($\xi \geq 0$ on Γ_4 would be enough but Γ_4 is unknown!) we obtain from (4.16)

$$\int_A \nabla p \cdot \nabla \xi + \xi_y \leq 0 \qquad \forall \xi \in C^1(\overline{\Omega}), \ \xi = 0 \text{ on } \Gamma_3, \ \xi \geq 0 \text{ on } S_2.$$

Hence p being equal to 0 outside A we can rewrite this inequality as

$$\int_\Omega \nabla p \cdot \nabla \xi + \chi(A)\xi_y \leq 0 \qquad \forall \xi \in C^1(\overline{\Omega}), \ \xi = 0 \text{ on } S_3, \ \xi \geq 0 \text{ on } S_2$$

where here and in the sequel $\chi(A)$ denotes the characteristic function of the set A. But now, by the maximum principle, A can be characterized as the set,

$$[p > 0] = \{(x,y) \in \Omega \mid p(x,y) > 0\}.$$

Indeed p is positive on $\Gamma_2 \cup \Gamma_3 \cup \Gamma_4$ and $\frac{\partial p}{\partial \nu} = -\nu_y$ on Γ_1 (see (4.15)). Since by our assumptions

$$\nu_y \leq 0 \text{ on } \Gamma_1, \tag{4.17}$$

the strong maximum principal gives us $p > 0$ inside A. Thus to look for a pair (p,A) or equivalently $(p,\chi(A))$ leads to considering the weaker problem:

(P) $\begin{cases} \text{Find a pair } (p,\chi) \in H^1(\Omega) \times L^\infty(\Omega) \text{ such that} \\[6pt] \text{(i) } p \geq 0 \text{ a.e. in } \Omega, \ p = \phi \text{ on } S_2 \cup S_3 \\[6pt] \text{(ii) } 0 \leq \chi \leq 1 \text{ a.e. in } \Omega, \ \chi = 1 \text{ a.e. on } [p > 0] \\[6pt] \text{(iii) } \int_\Omega \nabla p \cdot \nabla \xi + \chi \cdot \xi_y \leq 0 \quad \forall \xi \in H^1(\Omega), \ \xi = 0 \text{ on } S_3, \ \xi \geq 0 \text{ on } S_2. \end{cases}$

First let us prove that this problem has a solution.

4.1.4. Existence of a Solution

If $H_0(p)$ denotes the following maximal monotone graph:

$$H_0(p) = \begin{cases} 1 & \text{if } p > 0 \\ [0,1] & \text{if } p = 0 \\ 0 & \text{if } p < 0, \end{cases}$$

then, clearly, the line (ii) of (P) simply expresses that

$$\chi \in H_0(p).$$

Introducing $H_\varepsilon(p)$ defined for $\varepsilon > 0$ as:

$$H_\varepsilon(p) = \begin{cases} 1 & \text{if } p \geq \varepsilon \\ \dfrac{p}{\varepsilon} & \text{if } 0 \leq p \leq \varepsilon \\ 0 & \text{if } p \leq 0, \end{cases}$$

it is natural to look for p as the limit of p_ε where p_ε is the solution of

$$(P_\varepsilon) \quad \begin{cases} p_\varepsilon \in H^1(\Omega), \quad p_\varepsilon = \phi \quad \text{on} \quad S_2 \cup S_3 \\[2mm] \displaystyle\int_\Omega \nabla p_\varepsilon \cdot \nabla \xi + H_\varepsilon(p_\varepsilon) \cdot \xi_y = 0 \quad \forall \xi \in H^1(\Omega), \quad \xi = 0 \quad \text{on} \quad S_2 \cup S_3. \end{cases}$$

First, let us study the penalization (P_ε) of the problem (P) and prove:

__Theorem 4.1.__   ([35]) There exists a solution p_ε of (P_ε) .

__Proof:__ Let us consider the mapping F_ε which for $v \in L^2(\Omega)$ associates $u_\varepsilon = F_\varepsilon(v)$ the solution of the problem:

$$\begin{cases} u_\varepsilon = \phi \quad \text{on} \quad S_2 \cup S_3 \\[2mm] \displaystyle\int_\Omega \nabla u_\varepsilon \cdot \nabla \xi + H_\varepsilon(v) \cdot \xi_y = 0 \quad \forall \xi \in H^1(\Omega), \; \xi = 0 \quad \text{on} \quad S_2 \cup S_3. \end{cases} \tag{4.18}$$

(Existence and uniqueness of u_ε results from Corollary 1.13 and Poincare's inequality which is valid for functions which vanish only on $S_2 \cup S_3$.)

If we still denote by ϕ a function in $H^1(\Omega)$ which agrees with ϕ on $S_2 \cup S_3$ and if we substitute $\xi = u_\varepsilon - \phi$ in (4.18) we easily obtain since $H_\varepsilon(v)$ is bounded by 1

$$\int_{\Omega} |\nabla(u_\varepsilon - \phi)|^2 \leq C(\phi) \tag{4.19}$$

where $C(\phi)$ is a constant which depends on ϕ only. Thus u_ε is bounded in $H^1(\Omega)$ and from the compactness of the identity mapping from $H^1(\Omega)$ to $L^2(\Omega)$ (see Theorem 2.5) it follows that F_ε is completely continuous. Moreover, according to Poincaré inequality, we have

$$\int_{\Omega} |u_\varepsilon - \phi|^2 \leq K \int_{\Omega} |\nabla(u_\varepsilon - \phi)|^2 \leq K \cdot C(\phi).$$

So, for R large enough, F_ε maps the ball of center 0 and radius R (in $L^2(\Omega)$) into itself. Applying Corollary 1.2, we get the existence of p_ε as a fixed point of F_ε.

Moreover, we have:

Theorem 4.2. The solution p_ε of (P_ε) is unique, $p_\varepsilon \geq 0$ almost everywhere in Ω, and the mapping $\phi \to p_\varepsilon$ is nondecreasing (namely, if ϕ_1, ϕ_2 are two Lipschitz continuous functions on $S_2 \cup S_3$, if $\phi_2 \geq \phi_1$, then the corresponding solutions $p_\varepsilon^1, p_\varepsilon^2$ of (P_ε) satisfy $p_\varepsilon^2 \geq p_\varepsilon^1$ almost everywhere in Ω).

Proof: (As in the previous theorem the Lipschitz character of ϕ is only needed to insure the existence of an extension of ϕ to Ω. Thus a more general ϕ can be considered. However the physically relevant case requires ϕ to be defined by (4.13)).

If $p_\varepsilon^1, p_\varepsilon^2$ are two solutions of (P_ε) corresponding to boundary values ϕ_1 and ϕ_2, respectively, we set $q = q_\varepsilon = p_\varepsilon^1 - p_\varepsilon^2$. Setting for $\delta > 0$:

$$f_\delta(x) = \begin{cases} (1 - \dfrac{\delta}{x})^+ & \text{if } x \geq 0 \\[2mm] 0 & \text{if } x \leq 0 \end{cases}$$

we easily see that since $q \leq 0$ on $S_2 \cup S_3$, $\xi = f_\delta(q)$ is a function in $H^1(\Omega)$ which vanishes on $S_2 \cup S_3$. Moreover, we have (see Theorem 2.3)

$$\nabla\xi = f_\delta'(q) \cdot \nabla q = \delta \cdot \chi([q > \delta]) \cdot \nabla q / q^2. \tag{4.20}$$

From the equality satisfied by $p_\varepsilon^1, p_\varepsilon^2$ we then deduce:

$$\int_{\Omega} \nabla q \cdot \nabla\xi = \int_{\Omega} (H_\varepsilon(p_\varepsilon^2) - H_\varepsilon(p_\varepsilon^1))\xi_y,$$

and thus by (4.20)

$$\int_{[q>\delta]} \delta \cdot \frac{|\nabla q|^2}{q^2} = \int_{[q>\delta]} \delta \cdot (H_\epsilon(p_\epsilon^2) - H_\epsilon(p_\epsilon^1)) \frac{q_y}{q^2} .$$

Using the Lipschitz continuity of the H_ϵ this implies:

$$\int_{[q>\delta]} \frac{|\nabla q|^2}{q^2} \leq \frac{1}{\epsilon} \int_{[q>\delta]} \frac{|q_y|}{q} .$$

Thus by the Cauchy-Schwarz inequality, we obtain

$$\int_\Omega |\nabla \; Log(1 + \frac{(q-\delta)^+}{\delta})|^2 = \int_{[q>\delta]} \frac{|\nabla q|^2}{q^2} \leq \frac{|\Omega|}{\epsilon^2}$$

where $|\Omega|$ denotes the measure of Ω.

Applying now the Poincaré inequality, we get:

$$\int_\Omega Log(1 + \frac{(q-\delta)^+}{\delta})^2 \leq K \frac{|\Omega|}{\epsilon^2}$$

with K independent of δ. But when δ goes to zero we deduce from the above inequality

$$q = p_\epsilon^1 - p_\epsilon^2 \leq 0 \quad \text{a.e.} \quad \text{on} \quad \Omega. \tag{4.21}$$

Applying this inequality for $\phi = \phi_1 = \phi_2$ and $p_\epsilon^1, p_\epsilon^2$ two solutions of (p_ϵ) leads to the uniqueness of the solution of (p_ϵ). Monotonicity of $\phi \to p_\epsilon$ then follows from (4.21), and since $\phi \geq 0$, $p_\epsilon \geq 0$ (0 being the solution of (P_ϵ) when $\phi = 0$). This completes the proof.

Using this approximation we are able to prove:

Theorem 4.3. Under the above assumptions, there exists a pair (p,χ) which is a solution of (P).

Proof: From (4.19), p_ϵ being some u_ϵ, we deduce that p_ϵ is bounded in $H^1(\Omega)$ independently of ϵ. Thus (see Theorem 2.5) we can find a sequence $\epsilon_n \to 0$ such that:

$$p_{\epsilon_n} \longrightarrow p \quad \text{in} \quad H^1(\Omega)$$

$$p_{\epsilon_n} \to p \quad \text{in} \quad L^2(\Omega)$$

$$H_{\epsilon_n}(p_{\epsilon_n}) \longrightarrow \chi \quad \text{in} \quad L^2(\Omega)$$

($H_\epsilon(p_\epsilon)$ is bounded in $L^\infty(\Omega)$ and hence also in $L^2(\Omega)$).

Moreover, if in (P_{ϵ_n}) we choose $\xi \in \mathcal{D}(\Omega)$ we obtain:

$$\Delta p_{\varepsilon_n} = -H'_{\varepsilon_n}(p_{\varepsilon_n})(p_{\varepsilon_n})_y \in L^2(\Omega)$$

and by an extended Green's formula (see [15]) we easily have:

$$\int_\Omega \nabla p_{\varepsilon_n} \cdot \nabla \xi + H_{\varepsilon_n}(p_{\varepsilon_n})\xi_y = \int_{S_2} \frac{\partial p_{\varepsilon_n}}{\partial \nu} \cdot \xi \leq 0 \quad \forall \xi \in H^1(\Omega), \ \xi = 0 \quad \text{on} \quad S_3,$$

$$\xi \geq 0 \quad \text{on} \quad S_2.$$

(Note that $p_{\varepsilon_n} \geq 0$ and $p_{\varepsilon_n} = 0$ on S_2 imply $\dfrac{\partial p_{\varepsilon_n}}{\partial \nu} \leq 0$ on S_2.)

Letting $\varepsilon_n \to 0$, (iii) of (P) follows.

On the other hand, since the convex sets

$$K = \{v \in H^1(\Omega) \mid v \geq 0 \quad \text{a.e. in} \quad \Omega, \ v = \phi \quad \text{on} \quad S_2 \cup S_3\}$$

$$K' = \{v \in L^2(\Omega) \mid 0 \leq v \leq 1 \quad \text{a.e. in} \quad \Omega\}$$

are weakly closed in $H^1(\Omega)$ and $L^2(\Omega)$, respectively, we conclude that $(p,\chi) \in K \times K'$. This gives (i) and the first part of (ii) in (P).

Finally, on the set $[p > 0]$ we have almost everywhere (after extraction of a subsequence) $H_{\varepsilon_n}(p_{\varepsilon_n}) \to 1$. Thus by Lebesgue's Theorem $H_{\varepsilon_n}(p_{\varepsilon_n}) \to 1$ in $L^2([p > 0])$. From the fact that $H_{\varepsilon_n}(p_{\varepsilon_n}) \to \chi$ in $L^2(\Omega)$ and by uniqueness of the limit we deduce that $\chi = 1$ almost everywhere on $[p > 0]$. The theorem follows.

Remark 4.4. Except in our formulation (see 4.17) we have not used the assumptions (4.2)-(4.5). Thus the problems (P) and (P) make sense in a general Lipschitz domain Ω. In particular the Theorems 4.1 - 4.3 extend to this case. Note that in the sequel an important point will be to prove that χ is the characteristic function of the set $[p > 0]$.

We now want to point out some properties of a solution pair (p,χ) of (P).

4.2. Some Properties of (p,χ) Solution of (P).

Under the assumptions (4.2)-(4.5) we have:

Theorem 4.5. Let (p,χ) be a solution of (P) then:

(i) $\Delta p + \chi_y = 0$ in Ω

(ii) $\Delta p \geq 0$ in Ω

(iii) $\chi_y \leq 0$ and thus χ is a nonincreasing function of y in Ω.

Proof: For (i) it is enough to choose $\xi \in \mathscr{D}(\Omega)$ in (P). For (ii) we use a technique of [5]. Choose $\zeta \in \mathscr{D}(\Omega)$, $\zeta \geq 0$ and for $\epsilon > 0$ set

$$\xi = \min(\tfrac{p}{\epsilon}, \zeta).$$

Since $\phi \geq 0$ on $S_2 \cup S_3$, ξ vanishes on $S_2 \cup S_3$ and ξ and $-\xi$ are both test functions for (P). Thus substituting these two functions in (P) (iii) we get:

$$\int_\Omega \nabla p \cdot \nabla \min(\tfrac{p}{\epsilon}, \zeta) + \int_\Omega \chi \cdot [\min(\tfrac{p}{\epsilon}, \zeta)]_y = 0.$$

Since $\chi = 1$ a.e. on $[p > 0]$ and $[\min(\tfrac{p}{\epsilon}, \zeta)]_y = 0$ a.e. on $[p = 0]$ (see Remark 2.5) the above equality becomes:

$$\int_\Omega \nabla p \cdot \nabla \min(\tfrac{p}{\epsilon}, \zeta) + \int_\Omega [\min(\tfrac{p}{\epsilon}, \zeta)]_y = 0,$$

and thus:

$$\int_\Omega \nabla p \cdot \nabla \min(\tfrac{p}{\epsilon}, \zeta) = 0 \quad (\text{since} \quad \min(\tfrac{p}{\epsilon}, \zeta) = 0 \quad \text{on} \quad S).$$

Writing this equality as

$$\int_{[p > \epsilon\zeta]} \nabla p \cdot \nabla \zeta + \frac{1}{\epsilon} \int_{[p \leq \epsilon\zeta]} |\nabla p|^2 = 0$$

we get

$$\int_\Omega \chi([p > \epsilon\zeta]) \nabla p \cdot \nabla \zeta \leq 0.$$

Now using the fact $\chi([p > \epsilon\zeta]) \to \chi([p > 0])$ a.e. when $\epsilon \to 0$, we obtain, by letting ϵ go to 0 in the above inequality and applying the Lebesgue Theorem

$$\int_\Omega \chi(p > 0) \cdot \nabla p \nabla \zeta = \int_\Omega \nabla p \cdot \nabla \zeta \leq 0 \qquad \forall \zeta \in \mathscr{D}(\Omega), \quad \zeta \geq 0.$$

This is precisely (ii). From (i), (iii) is then obvious (see [98] and note that by our assumption on the shape of Ω, every vertical strip in Ω is connected).

Remark 4.6. In the case of a general domain (i), (ii) follow in the same way. Then, we can choose all at once a decomposition of Ω in $\Omega_1, \Omega_2, \ldots, \Omega_p$, (see (4.22)) each Ω_j vertically convex open subsets of Ω (see (4.2)). In this situation the part (iii) of the previous theorem become, $\chi_y \leq 0$ and χ is a nonincreasing function of y in each Ω_j.

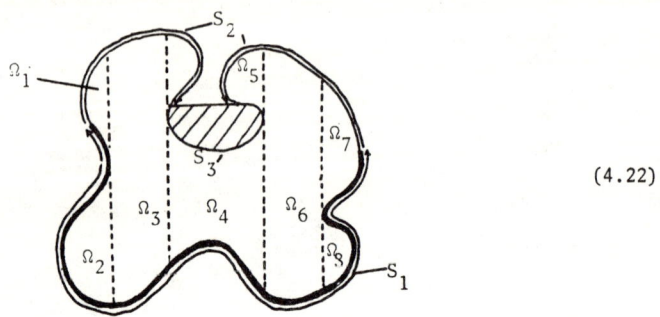

(4.22)

In the sequel, in the general case, we shall tacitly assume that such a decomposition of Ω into open sets Ω_j is possible. Similar to (4.6), (4.10), (4.11), setting for $x \in \pi_x(\Omega_j)$

$$S_j^-(x) = \mathrm{Inf}\{y \mid (x,y) \in \Omega_j\}, \quad S_j^+(x) = \mathrm{Sup}\{y \mid (x,y) \in \Omega_j\},$$

we shall assume that S_j^- and S_j^+ are continuous except on a finite set $\mathscr{S}_j^-$ and $\mathscr{S}_j^+$ respectively.

The next statement does not depend on the shape of Ω. We have:

Theorem 4.7. For all (p,χ) solution of (P), p is continuous on all Ω and on $S_2 \cup S_3$, thus the set $[p > 0]$ is open.

Proof: By localization near $S_2 \cup S_3$ this follows easily from (i) of Theorem 4.5 and from the usual regularity results (see for instance [65], p. 196).

Remark 4.8. Note that by Theorem 4.5 (i), p belongs to $W_{\mathrm{Loc}}^{1,s}(\Omega)$ for all $s > 1$ (see [97]). Usually $u \notin W^{1,\infty}(\Omega)$ (see [10], and the recent results of [8]).

Note also that (see Theorem 4.5 (i) and (P)(ii)) we have $\Delta p = 0$ in the open set $[p > 0]$. Thus p is analytic in $[p > 0]$. This will be used later.

Let us now point out the properties of the set $[p > 0]$. First:

Theorem 4.9. Let (p,χ) be a solution of (P). If $(x_0,y_0) \in [p > 0]$ then there exists an $\varepsilon > 0$ such that the cylinder

$$C_\varepsilon = \{(x,y) \in \Omega \mid |x-x_0| < \varepsilon, \ y < y_0 + \varepsilon\}$$

lies in $[p > 0]$.

Proof: If $(x_0,y_0) \in [p > 0]$ and since this set is open, for ε small enough the square

$$Q_\varepsilon = \{(x,y) \in \Omega |\ |x-x_0| < \varepsilon,\ |y-y_0| < \varepsilon\}$$

is also included in $[p > 0]$. Thus we have $\chi = 1$ on Q_ε and, by monotonicity of χ, $\chi = 1$ on C_ε (we have used here the fact that by (4.2), C_ε is connected by vertical segments). But from Theorem 4.5 (i) we now obtain $\Delta p = 0$ on C_ε, and if p takes the value 0 on C_ε by the maximum principle p is identically 0 on C_ε, which contradicts $Q_\varepsilon \subset [p > 0]$.

As a consequence we have:

Corollary 4.10. Let (p,χ) be a solution of (P). If $p(x_0,y_0) = 0$, then $p(x_0,y) = 0$ for all $(x_0,y) \in \Omega$ with $y \geq y_0$.

Proof: Otherwise, we would have $p(x_0,y) > 0$ for some $y > y_0$ and, from Theorem 4.9, a contradiction.

Remark 4.11. In the general case (see Remark 4.6) Theorems 3.9 and Corollary 4.10 are replaced by the following: choose a point $(x_0,y_0) \in [p > 0]$, then for ε small enough, we have $p > 0$ on each straight segment coming down, starting in Q_ε and included in Ω. If $(x_0,y_0) \in \Omega$ and $p(x_0,y_0) = 0$ then $p(x_0,y) = 0$ for all $(x_0,y) \in \Omega$ such that $y \geq y_0$ and $\{x_0\} \times [y_0,y] \subset \Omega$. The proof is the same as above.

Remark 4.12. Physically the meaning of Theorem 4.9 is clear: when a point (x_0,y_0) is wet, then by gravity acting, all is wet below. However note that this is no longer the case when the permeability is not constant (see [17]).

Concerning the wet set, under the assumptions (4.2)-(4.5) we can prove:

Theorem 4.13. Let (p,χ) be a solution of (P), then

$$p(x,y) > 0 \quad \forall (x,y) \in \Omega, \quad x \in \pi_x(S_3)$$

in other words below S_3 the dam is wet.

Proof: If $(x,y) \in S_3$, we have $p(x,y) = h_i - y > 0$. By Theorem 4.7, p is strictly positive in a neighborhood of (x,y) and the result follows from Theorem 4.9. (In the general case we have only $p > 0$ below S_3 up to the first point of $S_1 \cup S_2$ that we encounter by coming down along vertical lines.)

About the set $[p = 0]$ we have in complete generality:

<u>Theorem 4.14</u>. Let (p,χ) be a solution of (P). Denote by B_r an open
ball of center (x_0,y_0) and radius r included in Ω. If $p = 0$ in
B_r, then

$$p = \chi = 0 \quad \text{in} \quad B_r.$$

<u>Proof</u>: By Corollary 4.10 and Remark 4.11 we have $p = 0$ above B_r, i.e.,
on the part C of Ω of points connected to B_r by vertical segments.
Choose a strip $\Sigma = \ell \times (h,+\infty)$ included in $B_r \cup (x_0-r,x_0+r) \times (y_0,+\infty)$.
For $\alpha \in \mathcal{D}(\ell)$, $\alpha \geq 0$, $\xi = \chi(C) \cdot \alpha \cdot (y-h)^+$ is a test function for P. Thus
we get

$$\int_C \nabla p \cdot \nabla \xi + \chi \cdot \xi_y = \int_{C \cap \Sigma} \chi \cdot \alpha \leq 0.$$

Since $\alpha \geq 0$ is arbitrary in $\mathcal{D}(\ell)$ and $\chi \geq 0$, this implies clearly
$\chi = 0$ a.e. on $\Sigma \cap C$ and thus on C.

   At this stage, let us note that if (p,χ) is a solution of (P) we
are able to define the free boundary Φ of our problem. Indeed under
the assumptions (4.2)-(4.5) set for $x \in \pi_x(\Omega)$

$$\Phi(x) = \begin{cases} \text{Sup}\{y \mid p(x,y) > 0\} & \text{if this set is not empty} \\ s^-(x) & \text{otherwise.} \end{cases} \tag{4.23}$$

By Theorem 4.9, Corollary 4.10 this definition makes sense and we have:

<u>Theorem 4.15</u>. For all solutions (p,χ) of (P), the function Φ is lower
semi-continuous (l.s.c.) on $\pi_x(\Omega)$ except perhaps on $\mathscr{S}^-$ (see (4.10)).
Thus Φ is measurable and we have:

$$[p > 0] = \{(x,y) \in \Omega \mid y < \Phi(x)\} = [y < \Phi(x)]. \tag{4.24}$$

<u>Proof</u>: The l.s.c. of Φ is clear (by (4.10)) on the points which are
not in $\mathscr{S}^-$ and where $\Phi(x) = s^-(x)$. If now $x_0 \in \pi_x(\Omega)$ and $\Phi(x_0) > s^-(x_0)$ for $\varepsilon > 0$, let y_0 be such that $\Phi(x_0) > y_0 > \Phi(x_0) - \varepsilon$ with
$(x_0,y_0) \in \Omega$. By definition of Φ and Corollary 4.10 we have of course
$p(x_0,y_0) > 0$ and thus $p(x,y) > 0$ on a ball B_α of center (x_0,y_0)
and radius α. This means that for $x \in (x_0-\alpha,x_0+\alpha)$, $\Phi(x) \geq y_0 \geq \Phi(x_0)-\varepsilon$.
Hence the result follows since (4.24) is easy to check.

<u>Remark 4.16</u>. One of the advantages of assumptions (4.2)-(4.5) is that in
the decomposition of Ω into Ω_j (see Remark 4.6) we actually have only
one Ω_j, namely Ω itself. So if one wants to extend the definition of
Φ to the general case it is easily done by defining one Φ_j in each Ω_j.

The formula is the same as in (4.23) with now Ω_j in place of Ω, $S^-(x)$ being replaced by S_j^-. Under the assumptions of the Remark 4.6 the different Φ_j are l.s.c. on their domain of definition except perhaps on $\mathcal{S}_j^-$ $(j = 1, \ldots, p)$.

Let us now state a result which will play a fundamental role in the sequel:

Theorem 4.17. Let (p, χ) be a solution of (P), h a real number. Let C be a connected component of the set $[p > 0] \cap [y > h]$. Clearly $\pi_x(C) = (x_0, x_1)$ is an interval. Set

$$Z_h = \Omega \cap (x_0, x_1) \times (h, +\infty).$$

If $\overline{Z}_h \cap S_3 = \emptyset$ we have

$$\int_{Z_h} \chi p_y + \chi^2 \leq \int_{Z_h} p_y + \chi \leq 0. \tag{4.25}$$

(Here as in the following, we denote by a bar the closure in $\mathbb{R}^2$. The meaning of the set $[y > h]$ is obvious, i.e., $[y > h] = \{(x,y) \in \Omega | y > h\}$).

Proof: Let us assume that we have proved the following inequality for all $\zeta \in H^1(Z_h) \cap C(\overline{Z}_h)$, ζ positive, vanishing on $[y = h]$:

$$\int_{Z_h} \nabla p \cdot \nabla \zeta + \chi([p > 0])\zeta_y \leq \int_{(x_0, x_1)} \zeta(x, \Phi(x)) dx \tag{4.26}$$

and choose α_ε a function in $\mathcal{D}((x_0, x_1))$, between 0 and 1, and which is equal to 1 on $(x_0 + \varepsilon, x_1 - \varepsilon)$, $\varepsilon > 0$ being small. We have:

$$\int_{Z_h} p_y + \chi = \int_{Z_h} \nabla p \cdot \nabla(y - h) + \chi \cdot (y - h)_y$$

$$= \int_{Z_h} \nabla p \cdot \nabla[\alpha_\varepsilon(y-h)] + \chi \cdot [\alpha_\varepsilon(y-h)]_y + \int_{Z_h} \nabla p \cdot \nabla[(1-\alpha_\varepsilon)(y-h)]$$

$$+ \chi \cdot [(1-\alpha_\varepsilon)(y-h)]_y.$$

Since $\overline{Z}_h \cap S_3 = \emptyset$, clearly $\chi(Z_h)\alpha_\varepsilon \cdot (y - h)$ is a test function for (P) and the first integral in the above formula is negative. Applying (4.26) to the second leads to

$$\int_{Z_h} p_y + \chi \leq \int_{(x_0, x_1)} (1-\alpha_\varepsilon) \cdot (\Phi(x) - h) dx + \int_{Z_h} (\chi - \chi([p > 0]))(1-\alpha_\varepsilon).$$

Letting $\varepsilon \to 0$ the result follows by Lebesgue's Theorem, provided that (4.26) holds. (The first inequality is clear since $\chi^2 \leq \chi$ and

$\chi p_y = p_y$.) To prove (4.26) let us first remark that for $\zeta \in H^1(Z_h) \cap C(\bar{Z}_h)$, with $\zeta \geq 0$, and $\zeta = 0$ on $[y = h]$, the function

$$\xi = \chi(Z_h)\min(\frac{P}{\varepsilon}, \zeta) \qquad (\varepsilon > 0)$$

is a test function for (P). (Note that $p(x_i, y) = 0 \quad \forall(x_i, y) \in \Omega, \ y \geq h$, $i = 0, 1$. This follows from the definition of C and from Theorem 4.9.) Thus from (P)(iii) we get:

$$\frac{1}{\varepsilon}\int_{Z_h \cap [p \leq \varepsilon\zeta]} |\nabla p|^2 + \int_{Z_h \cap [p > \varepsilon\zeta]} \nabla p \cdot \nabla \zeta + \int_{Z_h} \chi \cdot [\min(\frac{P}{\varepsilon}, \zeta)]_y \leq 0.$$

Since $[\min(\frac{P}{\varepsilon}, \zeta)]_y = 0$ almost everywhere on $[p = 0]$ this leads to

$$\int_{Z_h} \chi([p > \varepsilon\zeta]) \nabla p \cdot \nabla \zeta + \chi([p > 0])[\min(\frac{P}{\varepsilon}, \zeta)]_y \leq 0$$

$$\Longleftrightarrow \int_{Z_h} \chi([p > \varepsilon\zeta]) \nabla p \cdot \nabla \zeta + \chi([p > 0])\zeta_y \leq \int_{Z_h} \chi([p > 0])[\zeta - \min(\frac{P}{\varepsilon}, \zeta)]_y$$

$$\Longleftrightarrow \int_{Z_h} \chi([p > \varepsilon\zeta]) \nabla p \cdot \nabla \zeta + \chi([p > 0])\zeta_y \leq \int_{Z_h} \chi([p > 0])(\zeta - \frac{P}{\varepsilon})_y^+. \qquad (4.27)$$

By Fubini's Theorem the last integral is equal to

$$\int_{(x_0, x_1)} \int_{(h, \Phi(x))} (\zeta - \frac{P}{\varepsilon})_y^+(x, y)dy \ dx,$$

where for simplicity we have still denoted by h the maximum of h and $S^-(x)$. (S_1 can indeed intersect $y = h$.) But now using the absolute continuity in y of $(\zeta - \frac{P}{\varepsilon})^+$ (see [98], p. 57) for almost all x in (x_0, x_1) such that $\Phi(x) > h$, and for δ small enough we have

$$\int_{[h+\delta, \Phi(x)-\delta]} (\zeta - \frac{P}{\varepsilon})_y^+(x, y)dy \leq (\zeta - \frac{P}{\varepsilon})^+(x, \Phi(x)-\delta) \leq \zeta(x, \Phi(x)-\delta).$$

Now letting δ go to zero, by continuity of ζ we get for almost all $x \in (x_0, x_1)$

$$\int_{(h, \Phi(x))} (\zeta - \frac{P}{\varepsilon})_y^+(x, y)dy \leq \zeta(x, \Phi(x))$$

and clearly (4.26) follows. (Note that in (4.26) the right-hand side makes sense by Theorem 4.15.)

<u>Remark 4.18</u>. In the general case of Remark 4.6 the result is as follows. As above let C be a connected component of $[p > 0] \cap [y > h]$.

If (Ω_j) denotes the decomposition of Ω in Ω_j, set:

$$\Omega'_j = \{(x,y) \in \Omega_j \,|\, x \in \pi_x(C \cap \Omega_j), \quad y > h\}$$

$$Z_h = \bigcup_j \Omega'_j.$$

(Note that each $\Omega'_j \neq \emptyset$ is a connected open set which is vertically convex. Roughly speaking Z_h is obtained as follows: pick (x,y) in C, draw the vertical segment from (x,y) up to the first points where it en-counters S, then Z_h is the union of such segments (x,y) running through C.) Now the result of Theorem 4.16 is the same: If $\overline{Z}_h \cap S_3 = \emptyset$, then (4.25) holds. First, with the same proof as above, (4.26) is replaced by:

$$\int_{Z_h} \nabla p \cdot \nabla \zeta + \chi([p > 0])\zeta_y \le \sum_{j=1}^{P} \int_{\pi_x(\Omega'_j)} \zeta(x, \Phi_j(x)) \, dx$$

for all $\zeta \in H^1(Z_h) \cap C(\overline{Z}_h)$, $\zeta \ge 0$, $\zeta = 0$ on $[y = h]$. (Φ_j is the free boundary defined in each Ω_j as in the Remark 4.15. The above sum is ob-tained by splitting (4.27) on the different Ω'_j.) Introducing then a function α_ε on each $\pi_x(\Omega'_j)$ the result follows as above. Once more the simplification under the assumptions (4.2)-(4.5) lies in the fact that Z_h is, in this case, comprised of only one Ω'_j.

Finally, to conclude this section let us prove the following result valid for any Ω:

Theorem 4.19. Let (p,χ) be a solution of (P), B_r an open ball of center (x_0,y_0) and radius r included in Ω. Then the following cannot occur:

(i) $\quad\begin{cases} p(x_0,y) = 0 & \forall (x_0,y) \in B_r \\ p(x,y) > 0 & \forall (x,y) \in B_r, \quad x \neq x_0 \end{cases}$

(ii) $\quad\begin{cases} p(x,y) = 0 & (x,y) \in B_r \cap [x \le x_0] \quad (\text{Resp. } B_r \cap [x \ge x_0]) \\ p(x,y) > 0 & (x,y) \in B_r \cap [x > x_0] \quad (\text{Resp. } B_r \cap [x < x_0]) \end{cases}$

Proof: Let $\xi \in \mathscr{D}(B_r)$. Since ξ and $-\xi$ are test functions for (P) we have:

$$\int_{B_r} \nabla p \cdot \nabla \xi + \chi \cdot \xi_y = 0. \tag{4.28}$$

Under the assumption (i) we have $\chi = 1$ a.e. on B_r and thus

$$\int_{B_r} \chi \cdot \xi_y = \int_{B_r} \xi_y = 0.$$

Under the assumption (ii) by Theorem 4.14 we have $\chi = 0$ a.e. on $B_r \cap$ $[x \leq x_0]$ and thus

$$\int_{B_r} \chi \cdot \xi_y = \int_{B_r \cap [x > x_0]} \xi_y = \int_{\partial(B_r \cap [x > x_0])} \xi \cdot \nu_y = 0.$$

(ν_y denotes the component in y of the outward unit normal $\vec{\nu}$ to $\partial(B_r \cap [x > x_0])$). Thus in both cases by (4.28)

$$\int_\Omega \nabla p \cdot \nabla \xi = 0 \qquad \xi \in \mathscr{D}(B_r) \Longleftrightarrow \Delta p = 0 \quad \text{in} \quad B_r.$$

But (i) and (ii) are now in contradiction with the maximum principle since p achieves its minimum inside B_r.

4.3. S_3-Connected Solutions

An important point, that we have not yet discussed is of course: does (P) have a unique solution? The answer is no. The simplest example is the following:

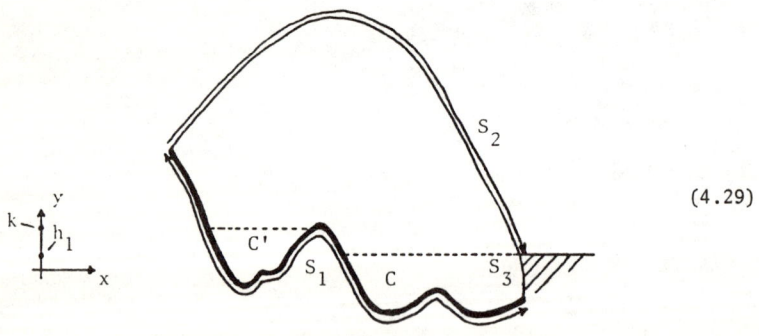

(4.29)

Assume that we are in the case of Figure (4.29) with C and C' denoting the regions indicated in (4.29). Then clearly

$$(p,\chi) = \begin{cases} (h_1 - y, 1) & \text{on} \quad C \\ (0,0) & \text{outside} \quad C \end{cases}$$

(4.30)

is a solution of (P), but so too is

$$(p,\chi) = \begin{cases} (h_1 - y, 1) & \text{on } C \\ (k - y, 1) & \text{on } C' \\ (0,0) & \text{elsewhere.} \end{cases}$$

Of course, the choice of C' is not unique.

One of our goals will now be to show that the Figure (4.29) provides us in fact with the only pathology which leads to nonuniqueness. More precisely (4.30) seems to be the only physically relevant solution of (P). It will be called, for obvious reasons, the S_3-connected solution of (P). Then we will show that all other solutions are obtained by adding to the S_3-connected solution pairs defined by $(k - y, 1)$ on sets of type C'. As we will see such a splitting of the solutions of (P) is quite general, no matter the shape of the dam considered.

Let us first begin with a precise definition.

<u>Definition 4.20</u>. We shall say that a solution (p,χ) of (P) is S_3-connected if for all connected components C of $[p > 0]$ we have

$$\overline{C} \cap S_3 \neq \emptyset.$$

(Recall that $\overline{C}$ denotes the closure of C in $\mathbb{R}^2$.)

<u>Remark 4.21</u>. Assume that $\overline{C} \cap S_{3,j} \neq \emptyset$ for some j. Then $\overline{C}$ contains a point P of $S_{3,j}$. Around this point (see Theorem 4.7) p is strictly positive. Thus C contains all the connected component of $[p > 0]$ which touches $S_{3,j}$. This shows also that if $\overline{C} \cap S_3 = \emptyset$ then C is wet without supply of water from the different reservoirs. This, of course, seems to be incorrect from a physical point of view.

Although the formulation (P) does not rule out completely the possibility of connected components of $[p > 0]$ which do not touch S_3, (see (4.29)) the shape of such a component as well as the pressure inside is easily described. Indeed, as claimed in the beginning of this section, we have:

<u>Theorem 4.22</u>. Let (p,χ) be a solution of (P) and C a connected component of $[p > 0]$ such that $\overline{C} \cap S_3 = \emptyset$. Then there is some h such that:

C is a connected component of $[y < h]$, $p = h - y$ on C.

<u>Proof</u>: Recall that $[y < h]$ is the set of points (x,y) in Ω such that $y < h$. Set $k = \text{Inf}\{y \mid (x,y) \in C\}$ and consider the set Z_k associated to C which is a connected component of $[p > 0] \cap [y > k]$ (the proof given here holds for a general dam, see Remark 4.18). By Theorem 4.17 we have

$$\int_{Z_k} \chi \cdot p_y + \chi^2 \leq 0.$$

On the other hand (see Remark 4.18) $p = 0$ on all boundary points of Z_k which are in Ω and $\pm \chi(Z_k)p$ is a test function for (P). (Recall that $\overline{Z}_k \cap S_3 = \emptyset$.) This leads to:

$$\int_{Z_k} |\nabla p|^2 + \chi \cdot p_y = 0.$$

Adding this to the above inequality we get

$$\int_{Z_k} p_x^2 + (\chi + p_y)^2 \leq 0.$$

Since $\chi = 1$ on $[p > 0]$ this leads to $\nabla p = (0,-1)$ on C and thus $p = (h - y)$ on C for some h. The result follows. Note that this implies that the bottom of such a C is surrounded by S_1 and that $(p,\chi) = 0$ in the remainder of Z_k.

Definition 4.23. Let us call a "pool" a pair of functions (p,χ) defined on a connected component of $[y < h]$ by

$$(p,\chi) = (h - y, 1).$$

Then we have:

Theorem 4.24. Any solution (p,χ) of (P) is the sum of an S_3-connected solution and "pools".

Proof: Let (p,χ) be a solution of (P). Denote by C_i the connected component of $[p > 0]$ such that $\overline{C}_i \cap S_3 = \emptyset$ ($i \in I$) and set

$$(p',\chi') = (p,\chi) - \sum_i (\chi(C_i)p, \chi(C_i)).$$

Then clearly all connected components C' of $[p' > 0]$ are such that $\overline{C}' \cap S_3 \neq \emptyset$. Moreover, by Theorem 4.22 we have

$$\int_\Omega \nabla p' \cdot \nabla \xi + \chi' \cdot \xi_y = \int_\Omega \nabla p \cdot \nabla \xi + \chi \cdot \xi_y - \sum_i \int_{C_i} -\xi_y + \xi_y \leq 0,$$

for all $\xi \in H^1(\Omega)$, $\xi \geq 0$ on S_2, $\xi = 0$ on S_3. Thus (p',χ') is a S_3-connected solution of (P) which concludes the proof.

This theorem allows us to restrict our attention to S_3-connected solutions.

First let us assume that the numbering of the reservoirs with bottom $S_{3,i}$ $(i = 1,...,n)$ is chosen so that

$$h_n \leq h_{n-1} \leq \cdots \leq h_1. \qquad (4.31)$$

Then the first natural result is to show that for an S_3-connected solution the level of the wet set cannot exceed the level h_1, and in fact, we have more precisely:

<u>Theorem 4.25.</u> Let (p,χ) be an S_3-connected solution of (P) then:

   (i) $p = \chi = 0$ in $[y > h_1]$

   (ii) $0 \leq p \leq (h_1 - y)^+$ in Ω.

<u>Proof</u>: Set $\xi = (p - (h_1-y)^+)^+$. Clearly ξ vanishes on $S_2 \cup S_3$ and by (P) (iii) we get:

$$\int_\Omega \nabla p \cdot \nabla(p - (h_1-y)^+))^+ + \chi \cdot (p - (h_1-y)^+)_y^+ \leq 0$$

$$\Longleftrightarrow \quad \int_{[y \leq h_1]} \nabla p \cdot \nabla(p - (h_1-y)^+)^+ + \chi \cdot (p - (h_1-y)^+)_y^+ + \int_{[y > h_1]} |\nabla p|^2 + \chi \cdot p_y \leq 0.$$

But on $[y \leq h_1]$ we have almost everywhere

$$\chi \cdot (p - (h_1-y)^+)_y^+ = \nabla - (h_1-y)^+ \cdot \nabla(p - (h_1-y)^+)^+.$$

(This results from the fact that $-(h_1-y)_x^+ = 0$, and since $\chi \cdot (p - (h_1-y)^+)_y^+$ is equal to $(p - (h_1-y)^+)_y^+ = -(h_1-y)_y^+(p - (h_1-y)^+)_y^+$ where $p > 0$ and to 0 when p vanishes.)

Thus the above inequality becomes:

$$\int_{[y \leq h_1]} |\nabla(p - (h_1-y)^+)^+|^2 + \int_{[y > h_1]} |\nabla p|^2 + \chi \cdot p_y \leq 0. \qquad (4.32)$$

In the second integral it is enough to integrate on the different connected components of $[p > 0] \cap [y > h_1]$ or still on the different $Z_{h_1}^i$ generated by these components (see Remark 4.18). But on these sets we have by Theorem 4.17

$$\int_{Z_{h_1}^i} \chi \cdot p_y + \chi^2 \leq 0.$$

Summing in i and adding to (4.26) leads to:

$$\int_{[y \leq h_1]} |\nabla(p - (h_1-y)^+)^+|^2 + \sum_i \int_{Z_{h_1}^i} p_x^2 + (p_y + \chi)^2 \leq 0.$$

Thus we deduce from this inequality that $\nabla p = (0,-1)$ on each connected component C^i of $[p > 0] \cap [y > h_1]$. This implies that $p = k - y$ on such a component, k being of course strictly greater than h_1. But by analytic continuation (see Remark 4.8) we must have $p = k - y$ on all the connected components of $[p > 0]$ containing C^i. Since (p,χ) is S_3-connected, such a component must touch S_3 which is impossible since $k > h_1$. Thus $p = 0$ on $[y > h_1]$, and (i) results from Theorem 4.14. (ii) results from the fact that (4.32) can now be written

$$\int_\Omega |\nabla(p - (h_1 - y)^+)^+|^2 \le 0.$$

Remark 4.26. The necessity of assuming that (p,χ) is S_3-connected is clear from the Figure (4.29).

Before describing more completely the properties of S_3-connected solutions let us take a look at the case of Baiocchi's dam, that is to say the rectangular one (see Figure (4.1)(B)).

In this case the main tool is to integrate along vertical lines (hence the necessity of vertical walls). More precisely, let (p,χ) be a solution of (P) in the case of the Figure (4.1)(B) set

$$u(x,y) = \int_y^{h_1} p(x,y)\,dy. \tag{4.33}$$

(This transformation is called the Baiocchi Transform), then we have:

Theorem 4.27. u defined by (4.33) satisfies:

(i) $[u > 0] = [p > 0]$

(ii) $-\Delta u + \chi = 0$ in Ω.

Proof: For (i), $u(x,y) > 0$ if and only if p is positive for some point (x,y') with $y' > y$. But by Theorem 4.9 it is the case if and only if $p(x,y) > 0$. For (ii) now, we first deduce from 4.33 that $p = -u_y$. Thus by Theorem 4.5(i)

$$(-\Delta u + \chi)_y = 0 \text{ in } \Omega$$

and $-\Delta u + \chi$ depends only on x. But on a level greater than h_1 (see (i) in the previous theorem) we have $p = u = \chi = 0$ and thus (ii) follows.

Remark 4.28. At this stage (ii) allows us to answer an important question (this technique applies to more general domains than rectangular ones but not in all generality). Indeed from regularity theory of elliptic problems we deduce from (i) that $u \in W^{2,s}_{loc}(\Omega)$ for all s, and thus (see

Remark 2.5) $\Delta u = \chi = 0$ almost everywhere on $[u = 0]$. This combined with (i) proves that χ is a characteristic function, i.e., that we have

$$\chi = \chi([p > 0]).\tag{4.34}$$

To obtain a variational inequality, it is enough to note that we have now

$$-\Delta u \geq -1, \quad u \geq 0, \quad (-\Delta u + 1)\cdot u = 0$$

(compare with 3.41). Moreover, 4.33 allows us to compute the boundary value of u. We have:

$$u = 0 \quad \text{on} \quad S_2$$

$$u = \int_y^{h_1}(h_1 - y)\,dy = \frac{(h_1 - y)^2}{2} \quad \text{on} \quad S_{3,1}\tag{4.35}$$

$$u = \int_y^{h_2}(h_2 - y)\,dy = \frac{(h_2 - y)^2}{2} \quad \text{on} \quad S_{3,2}.$$

Now to get the value of u on S_1, assume that the dam is wet around S_1 i.e., that $p > 0$ around S_1 (this can be shown easily). Then we have

$$\Delta u = 1$$

around S_1. Moreover, by (4.33) and (4.15)

$$u_{yy} = -p_y = 1 \quad \text{on} \quad S_1$$

and from $\Delta u = 1$ it follows that u_{xx} has to be equal to 0 on S_1 that is to say, u is linear on S_1. This leads with the notation of (4.1)(B) to

$$u = (a - x)\frac{h_1^2}{2a} + x\cdot\frac{h_2^2}{2a} \quad \text{on} \quad S_1.\tag{4.36}$$

Thus, introducing K as the set

$$K = \{u \in H^1(\Omega)\,|\,u \geq 0, \quad u \text{ satisfy (4.35), (4.36)}\}$$

u has to be a solution of

$$\begin{cases} <-\Delta u, v - u> \geq <-1, v - u> \quad \forall v \in K \\ \\ u \in K. \end{cases}$$

This problem was investigated for the first time by Baiocchi in [9], [10]. The uniqueness of u of course leads to the uniqueness of (p,χ) in this case. Note that the knowledge of u gives us p as $-u_y$ (see

[10], [15], [76]). For extensions of this method to higher dimensions the reader is referred to [76] and [102].

To prove (4.34) in all cases, as well as to establish some monotonicity results about the free boundary of our problem, the next two theorems will be very convenient tools. First we have in all generality:

<u>Theorem 4.29</u>. Let (p,χ) be a S_3-connected solution of (P). For all $k = 1,\ldots,n$ and all h satisfying $h_{k+1} \leq h < h_k$, the set

$$P_h = \{(x,y) \in \Omega \mid p(x,y) > (h - y)^+\}$$

has at most k connected components. More precisely, if for $i = 1,\ldots,k$ we denote by $C_{h,i}$ the connected component of P_h whose closure in $\mathbb{R}^2$ satisfies $\overline{C_{h,i}} \supset S_{3,i}$ one has:

$$P_h = C_{h,1} \cup C_{h,2} \cup \ldots \cup C_{h,k}. \tag{4.37}$$

(Of course in the above formula some of the $C_{h,i}$ can be the same. Note also that when $k = n$, h_{n+1} not being defined the assumption has simply to be read $h < h_n$. In the case $h \geq h_1$ it has already been proved that $P_h = \emptyset$. See Theorem 4.25.)

<u>Proof</u>: Set $C'_h = \Omega - \bigcup_{i=1}^{k} C_{h,i}$. Then

$$\xi = (1 - \sum_{i=1}^{k} \chi(C_{h,i})) \cdot (p - (h-y)^+)^+ = \chi(C'_h) \cdot (P - (h-y)^+)^+$$

is a test function for (P) (see [49] for a complete justification. Note that in the above sum the same $C_{h,i}$ are assumed to be taken into account once). Thus we get:

$$\int_{C'_h} \nabla p \cdot \nabla (p - (h-y)^+)^+ + \chi \cdot (p - (h-y)^+)^+_y \leq 0$$

$$\Longleftrightarrow \int_{C'_h \cap [y \leq h]} \nabla p \cdot \nabla (p - (h-y)^+)^+ + \chi \cdot (p - (h-y)^+)^+_y +$$

$$+ \int_{C'_h \cap [y > h]} |\nabla p|^2 + \chi \cdot p_y \leq 0.$$

But, (see the proof of Theorem 4.25) on $[y \leq h]$ one has

$$\chi \cdot (p - (h-y)^+)^+_y = \nabla - (h-y)^+ \cdot \nabla (p - (h-y)^+)^+,$$

and thus the above inequality becomes:

$$\int_{C_h' \cap [y \leq h]} |\nabla (p - (h-y)^+)^+|^2 + \int_{C_h' \cap [y > h]} |\nabla p|^2 + \chi \cdot p_y \leq 0. \qquad (4.38)$$

In the second integral it is clearly enough to integrate on the different connected components C^j of $[p > 0] \cap [y > h]$ which are not included in some $C_{h,i}$. Of course, this is still the same as integrating over the z_h^j generated by these connected components, (see Remark 4.18) and since such a z_h^j satisfies $\bar{z}_h^j \cap S_3 = \emptyset$ we have by Theorem 4.17

$$\int_{z_h^j} \chi \cdot p_y + \chi^2 \leq 0.$$

Combining this with (4.32) we get

$$\int_{C_h' \cap [y \leq h]} |\nabla (p - (h-y)^+)^+|^2 + \sum_j \int_{z_h^j} p_x^2 + (p_y + \chi)^2 \leq 0.$$

Thus on the connected components C^j we must have $p = k - y$, where $k > h$. By analytic continuation $p = k - y$ on all the connected component of $[p > 0]$ containing C_j. If (p,χ) is S_3-connected and C_j not included in some $C_{h,i}$ this is impossible. So all the C_j are empty and (4.38) becomes

$$\int_{C_h'} |\nabla (p - (h-y)^+)^+|^2 \leq 0.$$

Thus $p \leq (h - y)^+$ on C_h' and (4.37) follows.

Let us now point out an interesting feature of these $C_{h,i}$. Assume first that (4.2)-(4.5) holds. Then we have:

<u>Theorem 4.30</u>. Let (p,χ) be an S_3-connected solution of (P). Let (x_1,h) and (x_2,h) be two points in the same connected component $C_{h,i}$ of P_h. Then if the segment $[x_1,x_2] \times \{h\}$ does not intersect S_2 one has for $(x,y) \in \Omega$:

$$p(x,y) > 0 \quad \forall x \in [x_1,x_2], \quad \forall y < h.$$

<u>Proof</u>: Let us assume $p(x_0,y_0) = 0$ for some point below $[x_1,x_2] \times \{h\}$. By definition of P_h

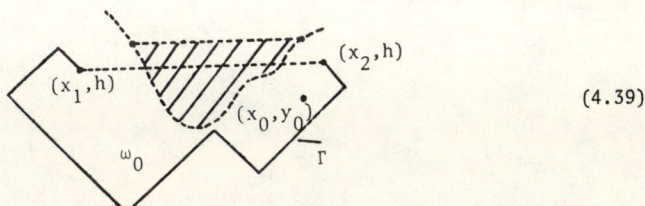

$$(4.39)$$

one has $p(x_i,h) > 0$. Thus, by Theorem 4.9, one can assume $(x_0,y_0) \in$
$(x_1,x_2) \times (-\infty,h)$. Now since $C_{h,i}$ is connected and open, it is arcwise
connected. Thus, one can find a path Γ joining (x_1,h) to (x_2,h)
(see Figure (4.39)) which is built with straight line segments and moreover
has no double point. Γ being in $C_{h,i}$ is in particular in $[p > 0]$.
Thus since $p(x_0,y) = 0$ for $(x_0,y) \in \Omega$, $y \geq y_0$ (see Corollary 4.10), Γ
has to cut the vertical line $x = x_0$ below (x_0,y_0) . Let us denote by
ω_0 the connected component of (x_0,y_0) in the complement of Γ in
$[y < h]$ and set:

$$\xi = -\chi(\omega_0)(p - (h - y))^-.$$

Clearly $\xi = 0$ on $[y = h]$ and in a neighborhood of Γ , by definition of P_h . Moreover, $\xi = 0$ on the $S_{3,j}$, which eventually intersects
$[x_1,x_2] \times \{h\}$ since for such a $S_{3,j}$ we have necessarily $h_j > h$. Thus,
ξ is a test function for (P) and we get

$$\int_{\omega_0} \nabla p \cdot \nabla[-(p - (h-y))^-] + \chi \cdot [-(p - (h-y))^-]_y \leq 0.$$

Since $\omega_0 \subset [y < h]$ this inequality becomes

$$\int_{\omega_0 \cap [p>0]} \nabla p \cdot \nabla[-(p - (h-y))^-] + \chi \cdot [-(p - (h-y))^-]_y + \int_{\omega_0 \cap [p=0]} \chi \leq 0.$$

Now since on $\omega_0 \cap [p > 0]$ we have $(0,\chi) = \nabla - [(h-y)]$ and since
$\chi \geq 0$ we have

$$\int_{\omega_0 \cap [p>0]} |\nabla(p - (h-y))^-|^2 \leq 0.$$

Thus $(p - (h-y))^-$ is constant on all connected components of
$\omega_0 \cap [p > 0]$. Let C be the connected component whose boundary contains
Γ . Since $p > (h - y)$ on Γ we have $p \geq h - y$ on C . But clearly C
is equal to ω_0 . Indeed otherwise C could have a boundary point in ω_0
and at this point we would have both $p = 0$ and $p \geq h - y > 0$. Thus we
have $p \geq h - y$ in ω_0 which contradicts $p(x_0,y_0) = 0$ and concludes
the proof.

Remark 4.31. The general case can be illustrated by the following pic-
ture:

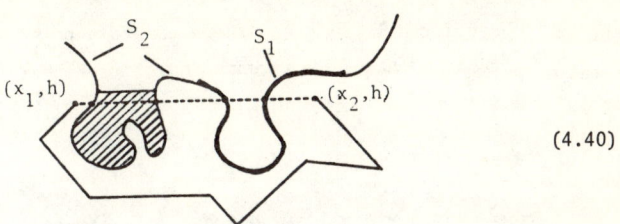

$$(4.40)$$

In this case clearly the result is the following (and the proof is the same as above): Let (p,χ) be an S_3-connected solution of (P) and (x_i,h); i = 1,2, be two points in the same $C_{h,i}$. If all connected components of $S \cap \mathbb{R} \times (-\infty,h)$ with end point on $[x_1,x_2] \times \{h\}$ are included in $S_1 \cup S_3$ then p is strictly positive below $[x_1,x_2] \times \{h\}$ up to the first point of S encountered by coming down along vertical lines. In particular, this is the case when (x_1,h), (x_2,h) are both in a ball included in Ω.

To conclude this section and assuming first that we are in the case (4.2)-(4.5) we have:

Theorem 4.32. Let (p,χ) be a solution of (P). The function Φ defined by (4.23) is continuous on $\pi_x(\Omega)$ except perhaps on $\mathscr{S}^- \cup \mathscr{S}^+$. Moreover Φ has right and left limits on $\mathscr{S}^+ \cup \mathscr{S}^-$ as well as a right or a left limit at the endpoints of $\pi_x(\Omega)$.

Proof: Clearly by Theorem 4.24 one can assume that (p,χ) is S_3-connected. Now let us first prove the following:

If $(x,y) \in \Omega$, $y > \Phi(x)$, then p vanishes in a
    neighborhood of (x,y).
$$(4.41)$$

Indeed assume that (4.41) fails. Let us denote by B_ε a ball included in Ω of center (x,y) and of radius $\varepsilon < y - \Phi(x)$. Two cases could occur:

(i) For all ball $B_{\varepsilon'}$ centered at (x,y) and of radius $\varepsilon' < \varepsilon$
     there exist two points $(\underline{x},\underline{y})$ and $(\overline{x},\overline{y})$ in $B_{\varepsilon'}$ such that

$$\underline{x} < x, \; p(\underline{x},\underline{y}) > 0 \quad \text{and} \quad x < \overline{x}, \; p(\overline{x},\overline{y}) > 0.$$

(ii) For ε' small enough, less than ε one has

$$p(\underline{x},\underline{y}) = 0 \quad \forall(\underline{x},\underline{y}) \in B_{\varepsilon'}, \quad \underline{x} < x$$

(Resp. $p(\overline{x},\overline{y}) = 0 \quad \forall(\overline{x},\overline{y}) \in B_{\varepsilon'}, \quad x < \overline{x})$

and for all $\varepsilon'' < \varepsilon'$ there exists a point $(\overline{x},\overline{y}) \in B_{\varepsilon''}$ (a ball
centered at (x,y)) such that

$$x < \overline{x}, \ P(\overline{x},\overline{y}) > 0$$

(Resp. $\underline{x} < x, \ P(\underline{x},\underline{y}) > 0$).

Let us consider for instance case (i). We thus can find two se-
quences (see Figure (4.42)).

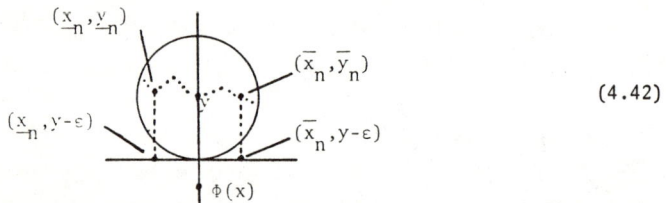

(4.42)

$(\underline{x}_n,\underline{y}_n)$ and $(\overline{x}_n,\overline{y}_n)$ in B_ε such that

$$\underline{x}_n < x, \ P(\underline{x}_n,\underline{y}_n) > 0, \ (\underline{x}_n,\underline{y}_n) \to (x,y) \ \text{when} \ n \to +\infty$$

$$x < \overline{x}_n, \ P(\overline{x}_n,\overline{y}_n) > 0, \ (\overline{x}_n,\overline{y}_n) \to (x,y) \ \text{when} \ n \to +\infty.$$

Moreover, we can assume $(\underline{x}_n)$ nondecreasing and $(\overline{x}_n)$ nonincreasing. Let
us then consider the sequence $(\underline{x}_n, y - \varepsilon)$. (If we choose ε small enough,
one can assume without loss of generality that this sequence as well as
the straight segment $[\underline{x}_1,x] \times \{y - \varepsilon\}$ are in Ω). From Theorem 4.9 we
deduce that $P(\underline{x}_n, y - \varepsilon) > 0$ for all n. Thus the points $(\underline{x}_n, y - \varepsilon)$
belong to $P_{y-\varepsilon}$ which (by Theorem 4.29) has only a finite number of con-
nected components. Thus there is an infinity of $(\underline{x}_n, y - \varepsilon)$ (starting
for instance from $(\underline{x}_1, y - \varepsilon)$) which belong to the same connected component of $P_{y-\varepsilon}$. By Theorem 4.30 this leads to $p > 0$ below $[\underline{x}_1,x) \times$
$\{y - \varepsilon\}$. But clearly the same applies on the other side of x and we
are in the case (i) of Theorem 4.19 (recall that $p(x,y') = 0$ for $y' \in$
$[\Phi(x),y - \varepsilon])$ and thus a contradiction. It is not difficult to see that
case (ii) would lead to case (ii) of Theorem 4.19, which proves (4.41) in
all cases.

To conclude the proof, let us first remark that by Theorem 4.15 we
already know that Φ is l.s.c. (lower semicontinuous) except perhaps on
$\mathscr{S}^-$. So let us prove that Φ is u.s.c. (upper semicontinuous) except
perhaps on $\mathscr{S}^+$. If we are on a point where $\Phi(x) = S^+(x)$ and $x \notin \mathscr{S}^+$
it is clear that Φ is u.s.c. since $\Phi \leq S^+$ and S^+ is continuous at x.

If now $\Phi(x) < S^+(x)$ the u.s.c. of Φ at x results from (4.41) (see the proof of Theorem 4.15). Thus Φ is continuous except perhaps on $\mathscr{S}^- \cup \mathscr{S}^+$.

To prove the existence of a left (or right) limit on a point of $\mathscr{S}^-$ or $\mathscr{S}^+$ (as well as the end of $\pi_x(\Omega)$) assume:

$$1 = \liminf_{\substack{x \to x_0 \\ x < x_0}} \Phi(x) < \limsup_{\substack{x \to x_0 \\ x < x_0}} \Phi(x) = L.$$

Then choose a sequence $x_n < x_0$ such that $\Phi(x_n) \to L$ when n goes to infinity. For ε small enough such that $L - \varepsilon > \ell$ we get by the same argument as above that an infinity of terms of the sequence $(x_n, L - \varepsilon)$ belongs to the same connected component of $P_{L-\varepsilon}$. Thus $p(x',y') > 0$ $\forall (x',y') \in \Omega$, $y' < L - \varepsilon$ for x' near x_0. This clearly implies that $\Phi(x') \geq L - \varepsilon > \ell$ near x_0 and a contradiction since then $\liminf_{\substack{x \to x_0 \\ x < x_0}} \Phi(x) \geq L - \varepsilon > \ell$.

Remark 4.33. In the general case, and with the same proof as above, the Theorem 4.32 has simply to be replaced by: Φ_j is continuous on each $\pi_x(\Omega_j)$ perhaps on $\mathscr{S}_j^- \cup \mathscr{S}_j^+$ (see Remark 4.6). Moreover, right and left limits exists on $\mathscr{S}_j^- \cup \mathscr{S}_j^+$ as well as right or left limit at the end points of $\pi_x(\Omega_j)$.

As a consequence we have now in complete generality:

Theorem 4.34. Let (p, χ) be a solution of (P), then

$$\chi = ([p > 0]). \tag{4.43}$$

Proof: This follows directly from the previous theorem and from Theorem 4.14 according to the fact that union of the sets $\{(x, \Phi_j(x)) \mid x \in \pi_x(\Omega_j)\}$ is of measure zero.

Let us now complete Theorem 4.24 by proving that there is only one S_3-connected solution of (P).

4.4. Uniqueness of S_3-Connected Solutions

This time let us argue directly in the general case. Let (p_1, χ_1), (p_2, χ_2) be two S_3-connected solutions of (P). Let us assume to be chosen a decomposition of Ω in Ω_j, $j = 1, \ldots, p$ (see Remark 4.6) and let us denote Φ_j^1, Φ_j^2 the function Φ_j (see Remark 4.16) corresponding to p_1, p_2 respectively in this decomposition.

Now set

$$P_0 = \min(P_1, P_2), \qquad \chi_0 = \min(\chi_1, \chi_2), \qquad \Phi_j^0 = \min(\Phi_j^1, \Phi_j^2)$$

the functions Φ_j^0 being defined in $\pi_x(\Omega_j)$, $j = 1, \ldots, p$.

First we have:

<u>Lemma 4.35.</u> For all $\zeta \in H^1(\Omega) \cap C(\overline{\Omega})$, $\zeta \geq 0$ we have for $i = 1$ or 2

$$\int_\Omega \nabla(P_i - P_0) \cdot \nabla\zeta + (\chi_i - \chi_0) \cdot \zeta_y \leq \sum_{j=1}^p \int_{D_j^i} \zeta(x, \Phi_j(x)) dx \qquad (4.44)$$

where

$$D_j^i = \{x \in \pi_x(\Omega_j) \mid \Phi_j^0(x) < \Phi_j^i(x)\}.$$

<u>Proof:</u> First note that the sets D_j^i are clearly measurable and the inte-
grals on the right side of (4.44) make sense. Let us give the proof for
instance for $i = 1$. Thus for ζ as above and $\varepsilon > 0$ consider

$$\xi = \min(\frac{P_1 - P_0}{\varepsilon}, \zeta).$$

Clearly $\xi = 0$ on $S_2 \cup S_3$ and $\pm\xi$ is a test function for (P).
Applying (P) (iii) corresponding respectively to P_1 and P_2 we get by
subtraction:

$$\int_\Omega \nabla(P_1 - P_2)\nabla\xi + (\chi_1 - \chi_2)\xi_y = 0.$$

But in this integral it is enough to integrate on the set $[P_1 - P_0 > 0]$
where $P_2 = P_0$, thus we have

$$\int_\Omega \nabla(P_1 - P_0) \cdot \nabla\xi + (\chi_1 - \chi_0) \cdot \xi_y = 0$$

$$\Longleftrightarrow \frac{1}{\varepsilon}\int_{[P_1 - P_0 \leq \varepsilon\zeta]} |\nabla(P_1 - P_0)|^2 + \int_{[P_1 - P_0 > \varepsilon\zeta]} \nabla(P_1 - P_0) \cdot \nabla\zeta \leq$$

$$+ \int_\Omega (\chi_1 - \chi_0) \cdot [\min(\frac{P_1 - P_0}{\varepsilon}, \zeta)]_y = 0.$$

This implies that

$$\int_{[P_1 - P_0 > \varepsilon\zeta]} \nabla(P_1 - P_0) \cdot \nabla\zeta + \int_\Omega (\chi_1 - \chi_0) \cdot \zeta_y \leq$$

$$\int_\Omega (\chi_1 - \chi_0) [\zeta - \min(\frac{P_1 - P_0}{\varepsilon}, \zeta)]_y.$$

But now splitting Ω into the different Ω_j in the integral on the right hand side we get (note that on $[p_0 > 0]$, $\chi_1 = \chi_0 = 1$ and use (4.43))

$$\int_\Omega (\chi_1 - \chi_0)[\zeta - \min(\frac{p_1 - p_0}{\varepsilon}, \zeta)]_y = \sum_{j=1}^{p} \int_{[\Phi_j^0(x) \leq y < \Phi_j^1(x)]} (\zeta - \frac{p_1}{\varepsilon})_y^+ \quad (4.45)$$

where

$$[\Phi_j^0(x) \leq y < \Phi_j^1(x)] = \{(x,y) \in \Omega \,|\, \Phi_j^0(x) \leq y < \Phi_j^1(x)\}.$$

Now, using the same arguments as in the conclusion of the proof of (4.26), i.e., integrating in y on the right hand side of (4.45) we get

$$\int_{[p_1 - p_0 > \varepsilon\zeta]} \nabla(p_1 - p_0) \cdot \nabla\zeta + \int_\Omega (\chi_1 - \chi_0) \cdot \zeta_y \leq \sum_{j=1}^{p} \int_{D_j^1} \zeta(x, \Phi_j^1(x))\,dx$$

and the lemma follows by letting ε go to zero.

We are now able to prove:

<u>Theorem 4.36</u>. There is one and only one S_3-connected solution of (P).

<u>Proof</u>: The existence results obviously from Theorem 2.42. The notation is always that of the preceding lemma. Choose B_r a ball centered on $S_{3,k}$ for some k such that B_r is included both in the connected component of $[p_1 > 0]$ and in those of $[p_2 > 0]$ touching $S_{3,k}$. (See Theorems 4.7 and 4.13 and the figure below.)

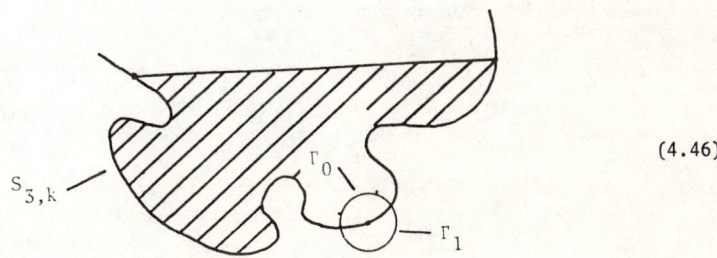

(4.46)

Let Γ_0 be an open part of ∂B_r (the boundary of B_r) outside Ω and Γ_1 the complement of Γ_0 in ∂B_r. Moreover, let us denote by σ a smooth function such that

$$\Delta\sigma = 0 \text{ in } B_r, \quad \sigma = 1 \text{ on } \Gamma_1, \quad 0 \leq \sigma < 1 \text{ on } \Gamma_0. \quad (4.47)$$

By Green's formula we have for $i = 1$ or 2

$$\int_{\Gamma_1 \cap \Omega} (p_i - p_0) \cdot \frac{\partial \sigma}{\partial \nu} = \int_{\Omega \cap B_r} \nabla(p_i - p_0) \cdot \nabla \sigma + \int_{\Omega \cap B_r} (p_i - p_0) \cdot \Delta \sigma$$

$$= \int_{\Omega \cap B_r} \nabla(p_i - p_0) \cdot \nabla \sigma.$$

If we still denote by σ the function (clearly in $H^1(\Omega) \cap C(\overline{\Omega})$) which agrees with σ in B_r and is equal to 1 outside, we get using the fact that $\chi_i = \chi_0 = 1$ on B_r:

$$\int_{\Gamma_1 \cap \Omega} (p_i - p_0) \cdot \frac{\partial \sigma}{\partial \nu} = \int_{\Omega} \nabla(p_i - p_0) \cdot \nabla \sigma + (\chi_1 - \chi_0) \cdot \sigma_y. \qquad (4.48)$$

Let us now denote by A_0 the set $[p_0 > 0]$. For $\varepsilon > 0$ let α_ε be a smooth function in $\mathbb{R}^2$ satisfying $0 \le \alpha_\varepsilon \le 1$, $\alpha_\varepsilon = 1$ on A_0, $\alpha_\varepsilon(x) = 0$ when $d(x, A_0) > \varepsilon$ (for instance $(1 - d(x, A_0)/\varepsilon)^+$ would be suitable.) By the maximum principle we deduce from (4.47) that $0 \le \sigma \le 1$. Thus $(1 - \alpha_\varepsilon)\sigma$ is a test function for (P) since it is a positive function which vanishes on A_0. This leads to:

$$\int_{\Omega} \nabla p_i \cdot \nabla[(1 - \alpha_\varepsilon)\sigma] + \chi_i \cdot [(1 - \alpha_\varepsilon)\sigma]_y \le 0 \qquad (i = 1,2).$$

Since $(1 - \alpha_\varepsilon)\sigma = 0$ on A_0 and also $p_0 = \chi_0 = 0$ outside A_0 we have:

$$\int_{\Omega} \nabla p_0 \cdot \nabla[(1 - \alpha_\varepsilon)\sigma] + \chi_0 \cdot [(1 - \alpha_\varepsilon)\sigma]_y = 0.$$

By subtracting from the above inequality we thus get:

$$\int_{\Omega} \nabla(p_i - p_0) \cdot \nabla \sigma + (\chi_i - \chi_0) \cdot \sigma_y \le \int_{\Omega} \nabla(p_i - p_0) \cdot \nabla(\alpha_\varepsilon \sigma) + (\chi_i - \chi_0)(\alpha_\varepsilon \sigma)_y.$$

Thus using (4.48) and (4.44) we obtain:

$$\int_{\Gamma_1 \cap \Omega} (p_i - p_0) \frac{\partial \sigma}{\partial \nu} \le \sum_{j=1}^{p} \int_{D_j^i} (\alpha_\varepsilon \sigma)(x, \phi_j^i(x)) dx \qquad (i = 1,2).$$

Now letting $\varepsilon \to 0$, we have by Lebesgue's Theorem and since $\alpha_\varepsilon(x, \phi_j^i(x)) \to 0$ on D_j^i

$$\int_{\Gamma_1 \cap \Omega} (p_i - p_0) \cdot \frac{\partial \sigma}{\partial \nu} \le 0 \qquad (i = 1,2).$$

But (4.47) and the maximum principle imply $\frac{\partial \sigma}{\partial \nu} > 0$ on Γ_1. So the above inequality implies $p_i = p_0$ on $\Gamma_1 \cap \Omega$ for $i = 1,2$ and

$$p_1 = p_2 \quad \text{on} \quad \Gamma_1 \cap \Omega.$$

Repeating the procedure for balls included in B_r, we get $p_1 = p_2$ in $B_r \cap \Omega$. By analytic continuation we obtain that $p_1 = p_2$ in C, the connected component of $[p_1 > 0] \cap [p_2 > 0]$ which contains $B_r \cap \Omega$. And so we easily obtain that

$$C = C_1 = C_2$$

where C_i denote the connected component of $[p_i > 0]$ which contains $S_{3,k}$ in its boundary. Indeed, $C \subset C_1$ and is an open nonempty subset of C_1. Now if $x_n \to x$ in C_1 from $p_1(x_n) = p_2(x_n)$ we deduce that $p_1(x) = p_2(x) > 0$ and thus C is closed in C_1. Hence $C = C_1 = C_2$ and the result follows since we can do the same proof for all $S_{3,k}$, $k = 1,\ldots,n$.

Remark 4.37.   One should note that the S_3-connected solution (p,χ) of (P) is also the minimal one, i.e., any other solution (p',χ') satisfies $p' \geq p$, $\chi' \geq \chi$ (see Theorem 2.24). It coincides also with the minimal solution of [5] (see [54]). Moreover, when the shape of Ω is such that no pools can appear (see Theorem 2.24) then clearly the solution of (P) is unique. This is the case for the rectangular dam and more generally for a dam with horizontal bottom as studied in [11] (see Figure (4.49)(A)). We have drawn on the Figure (4.49)(B) a case of a more complicated Ω where such uniqueness holds too, since no pool can appear!

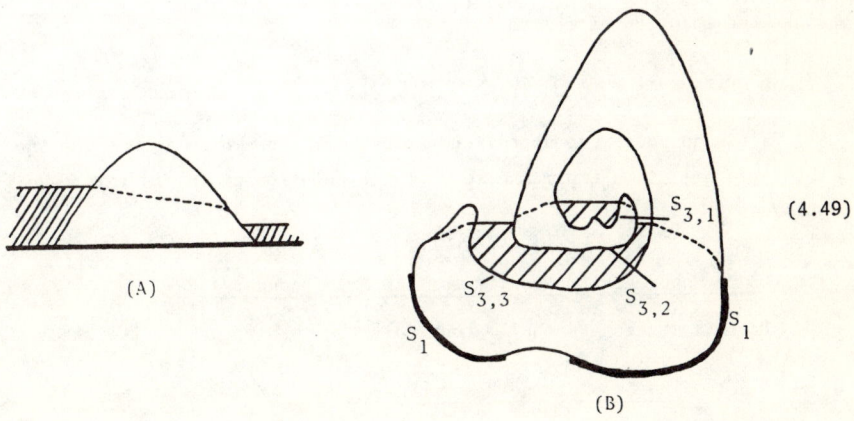

(A)   (B)   (4.49)

The reader is referred to [49] for more results in this direction.

Let us conclude this section by a theorem which is useful in deciding what part of the dam is wet as well as whether uniqueness holds or not. Physically, the result is simply that the pressure of the S_3-connected solution increases if the levels of the reservoirs increase.

Theorem 4.38. Let (p_1, χ_1), (p_2, χ_2) be two S_3-connected solutions of (P) associated respectively to the functions ϕ_1 and ϕ_2 as in (4.13). (Obviously S_1 is assumed to be prescribed and S_3, S_2 are defined by the data of ϕ_1, ϕ_2.) Then if $\phi_2 \geq \phi_1$ we have $p_2 \geq p_1$, $\chi_2 \geq \chi_1$.

Proof: From (4.43) it is obviously enough to prove that $p_2 \geq p_1$. If we denote by $p_1^\varepsilon, p_2^\varepsilon$ the solutions of (P_ε) (see Section 4.1) associated with ϕ_1, ϕ_2 respectively, by Theorem 4.2 we have $p_2^\varepsilon \geq p_1^\varepsilon$. Now (see Theorem 4.3) we can find a sequence $\varepsilon_n \to 0$ such that $p_i^{\varepsilon_n} \to p_i'$ in $L^2(\Omega)$ $(i = 1,2)$. Thus letting ε_n go to zero in $p_2^{\varepsilon_n} \geq p_1^{\varepsilon_n}$ we get $p_2' \geq p_1' \geq p_1$ (since the S_3-connected solution is the minimal one). But this implies clearly that $p_2 \geq p_1$ since on the connected components of $[p_1 > 0]$ we have $p_2' > 0$ and thus $p_2' = p_2$. (Note that such a component is connected to S_3 in both cases $i = 1,2$.)

Remark 4.39. As a consequence of this theorem consider for instance a dam as in the Figure (4.1)(A). If we assume the above reservoir empty (i.e., $S_3 = S_{3,2}$), then clearly the S_3-connected solution of (P) in this situation is given by $((h_2 - y)^+, \chi([y < h_2]))$. Thus, if we assume now that both reservoirs are full, then the Theorem 4.38 implies that the S_3-connected solution of (P) satisfies

$$p \geq (h_2 - y)^+.$$

Since no pools can appear on the part of S_1 which is outside the set $[y < h_2]$ this implies also that the solution of (P) in this case is unique.

4.5. Some Monotonicity Results for the Free Boundary

In this section we shall study only the case of one or two reservoirs.

4.5.1. The Case of One Reservoir

The notation will be that of (4.50).

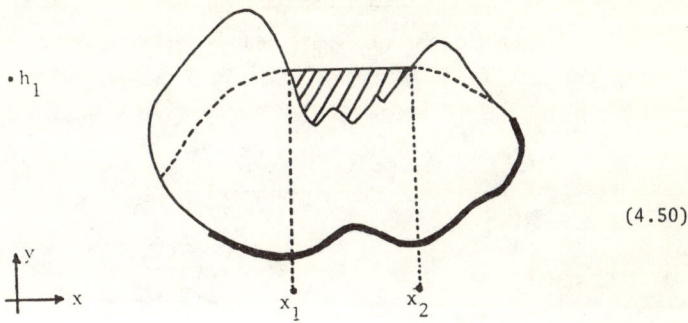

$$(4.50)$$

Ω is here a general domain satisfying, for simplicity, (4.2) and

For all (x,y), $(x',y) \in \Omega$ such that $y < h_1$

we have $(x,x') \times \{y\} \cap S_2 = \emptyset$.

$$(4.51)$$

Then we have:

Theorem 4.40. Let (p,χ) be the S_3-connected solution of (P) and Φ the function defined by (4.23). Then under the assumption (4.51) we have:

Φ is a non-decreasing function on the interval $(-\infty,x_1) \cap \pi_x([p > 0])$

Φ is a non-increasing function on the interval $(x_2,+\infty) \cap \pi_x([p > 0])$.

Proof: Let $x \in (x_2,+\infty) \cap \pi_x([p > 0])$ and $h < \Phi(x) \leq h_1$. If there exists a point $(x_0,y_0) \in \Omega$ which is below $(x_2,x) \times \{h\}$, and such that $p(x_0,y_0) = 0$, we would have in Ω, $p(x_0,y) = 0$ $\forall y \geq y_0$. Now by Theorem 4.7 one can find a point (α,β) in Ω near S_3 such that $\alpha < x_0$, $\beta > h$, $p(\alpha,\beta) > 0$. Clearly this point is in P_h which is connected (see Theorem 4.29). So since (x,h) is also in P_h and since $p(x_0,y) = 0$ $\forall y \geq y_0$, an arc connecting (x,h) to (α,β) must intersect in Ω the half line $(-\infty,x_0) \times \{h\}$ at some point which would be thus in P_h. But using (4.51), Theorem 4.30 and Remark 4.31 this leads to a contradiction of $p(x_0,y_0) = 0$. Thus in Ω below $(x_2,x) \times \{h\}$ we have $p > 0$ and $\Phi(x') \geq h$ for $x' \in (x_2,x)$. Since h is arbitrary we have $\Phi(x') \geq \Phi(x)$ and the result follows as the proof is the same on the other side.

Remark 4.41. In fact, except for some trivial cases where p is given by $(h_1 - y)^+$, the free boundary can never be stationary in Ω. This fact is also true for more than one reservoir. A proof is given in [49]. Note that this leads to some nontrivial examples where the uniqueness of the solution of (P) fails. (See also [49].)

4.5.2.   The Case of Two Reservoirs

The notation is that of (4.52) and we assume that (4.2), (4.51) hold.

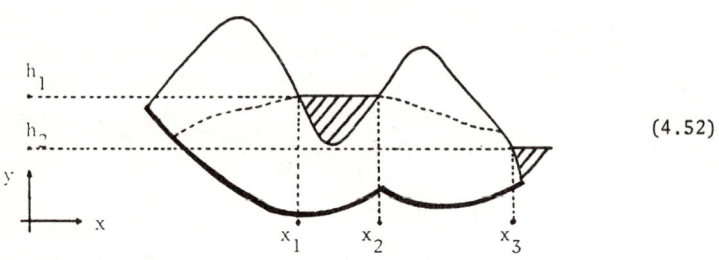

$$(4.52)$$

__Theorem 4.42.__   Let (p,χ) be the S_3-connected solution of (P) and the function defined by (4.23).   Then under the assumption (4.51) we have:

(i) Φ is a non-decreasing function of the interval $(-\infty,x_1)$ $\cap$
$\pi_x([p > 0])$.

Moreover, if we assume $[p > 0]$ is connected (otherwise it would be the case of one reservoir) then:

(ii)   If $\Phi(x) \geq h_2$ for all $x \in (x_2,x_3)$ then Φ is non-increasing on (x_2,x_3) (see 4.52).

(iii)   If there exists an $x \in (x_2,x_3)$ such that $\Phi(x) < h_2$, then there exists $x_m \in (x_2,x_3)$ such that $\Phi(x_m) < h_2$ and Φ is non-increasing on (x_2,x_m), Φ is non-decreasing on (x_m,x_3).

__Proof:__   Here for $h < h_i$ we shall denote by (α_i,β_i) a point near an end point of $S_{3,i}$ and which satisfies $p(\alpha_i,\beta_i) > 0$, that is to say $(\alpha_i,\beta_i) \in C_{h,i}$.   (Clearly such a point exists by Theorem 4.7.)
First let $x < x_1$ and $h < \Phi(x) \leq h_1$.   If there exists a point $(x_0,y_0) \in \Omega$ below $(x,x_1) \times \{h\}$ such that $p(x_0,y_0) = 0$, then we would have in Ω, $p(x_0,y) = 0$ $\forall y \geq y_0$.   Clearly by definition of Φ we have $(x,h) \in P_h$.   Assume $(x,h) \in C_{h,1}$ and (α_1,β_1) as above with moreover $\alpha_1 > x_0$.   Then any arc binding (x,h) to (α_1,β_1) must intersect in Ω the half line $(x_0,+\infty) \times \{h\}$ at some point.   But by (4.51), Theorem 4.30 and Remark 4.31 this contradicts $p(x_0,y_0) = 0$.   By assuming $(x,h) \in C_{h,2}$ we would necessarily have $h < h_2$ and a contradiction would be obtained by considering the arc connecting (x,h) to (α_2,β_2).   Thus we have $p > 0$ below $[x,x_1) \times \{h\}$ and consequently $\Phi(x') \geq h$, $\forall x' \geq x$, but h

being arbitrary leads to $\Phi(x') \geq \Phi(x)$ and to (i). Clearly (ii) holds
with the same proof by considering a point x such that $\Phi(x) > h_2$. Now
to prove (iii) let us denote by x_m a point in (x_2, x_3) such that

$$\Phi(x_m) = \inf_{x \in (x_2, x_3)} \Phi(x) < h_2.$$

Such a point exists by Theorem 4.15. Let us prove then that, for instance,
Φ is nondecreasing on the interval $[x_m, x_3)$. Let $x \in (x_m, x_3)$. If
$\Phi(x) = \Phi(x_m)$, then clearly by definition of x_m, $\Phi(x') \geq \Phi(x)$ $\forall x' \geq x$,
so let us assume $\Phi(x_m) < \Phi(x)$ and consider h such that $\Phi(x_m) < h <$
$\Phi(x)$. First if $(x,h) \in C_{h,1}$ since in Ω we have $p(x_m, y) = 0$ $\forall y \geq h$
the arc connecting (x,h) to (α_1, β_1) with $\alpha_1 < x_m$ must intersect the
half line $(-\infty, x_m] \times \{h\}$ at some point. But by (4.51) and Theorem 4.30
we would have a contradiction with $p(x_m, \Phi(x_m)) = 0$, thus $(x,h) \in C_{h,2}$,
which implies in particular $h < h_2$. By considering as previously an arc
connected (x,h) to some (α_2, β_2) we deduce then that $\Phi(x') \geq h$
$\forall x' > x$ and the proof is complete since h is arbitrary.

Remark 4.43. The Figure (4.53) gives an example where (iii) occurs.

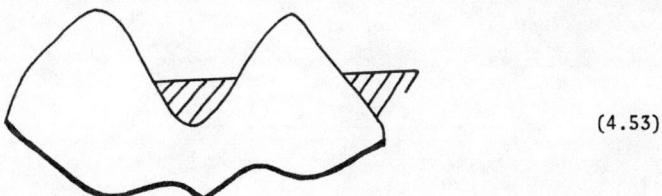

(4.53)

Here $h_1 = h_2$ and thus $\Phi(x) \leq h_1$. But if between the reservoirs
$\Phi(x) \equiv h_1$ we have a contradiction with the Remark 4.41. Thus we are in
case (iii), see [49].

Comments

The presentation here follows Brezis [32] and Carrillo-Menendez-Chipot [49]. When the problem reduces to a variational inequality a nice exposition is given in Kinderlehrer-Stampacchia [76]. See also Baiocchi-Capelo [15]. Other approaches are given in Alt [5] and Visintin [109]. Recently Alt and Gilardi [8] studied the manner in which the free boundary meets the fixed boundary of the domain; they give also other proofs of some of our results.

Most of the matter exposed here has an extension to $\mathbb{R}^p$; $p = 3$ is, of course, the physical case, the dimension 2 being a good model for flow in canals, Ω being the plane section of the porous medium (see [49]). Finally, we did not study the evolution problem. Details on this subject as well as complete references can be found in the proceedings [13].

Our model studies the case where the pressure is prescribed on the pervious boundary. Some other models are also valuable. In particular, the interested reader will find in [13] the derivation of such a model from the macroscopic point of view.

We note in conclusion that the prescription of p on the pervious boundary leads to realistic prediction on location and shape of the wet set (see Theorem 4.38, Section 4.5 and [49]). This should be of interest in hydrodynamics.

References

[1] R. A. Adams: Sobolev Spaces, Academic Press, New York (1975).

[2] D. R. Adams: L^p-capacitary integrals with some applications -
Proc. Symp. Pure Math. 35, Part I - A.M.S., Providence (1979), 359-
367.

[3] D. R. Adams: Capacity and the obstacle problem. Appl. Math. Optim.
8 (1981), 39-57.

[4] S. Agmon, A. Douglis and L. Nirenberg: Estimates near the boundary
for solution of elliptic partial differential equation satisfying
general boundary conditions. Comm. Pure Appl. Math. 12 (1959),
623-727.

[5] H. W. Alt: A free boundary problem associated with the flow of
ground water. Arch. Rat. Mech. Anal. 64 (1977), 111-126.

[6] H. W. Alt: The fluid flow through porous media. Regularity of
the free boundary. Manuscripta Math. 21 (1977), 255-272.

[7] H. W. Alt: Strömungen durch inhomogene poröse medien mit freiem
Rand. J. Reine Angew. Math. 305 (1979), 89-115.

[8] H. W. Alt and G. Gilardi: The behavior of the free boundary for the
dam problem, preprint.

[9] C. Baiocchi: Sur un problème à frontière libre traduisant le fil-
trage de liquides à travers des milieux poreux. C.R. Acad. Sc. Paris
Série A 273 (1971), 1215-1217.

[10] C. Baiocchi: Su un problema di frontiera libera connesso a ques-
tioni di idraulica. Ann. Mat. Pura Appl. 92 (1972), 107-127.

[11] C. Baiocchi: Free boundary problems in the theory of fluid flow
through porous media, Proc. Intern. Congress Math. Vancouver (1974)
II, 237-243.

[12] C. Baiocchi: Studio di un problema quasi-variazionale conesso a
problemi di frontiera libera. Boll. U.M.I. (4) 11 (Suppl. fasc. 3),
(1975), 589-613.

[13] C. Baiocchi: Free boundary problems in fluid flow through porous
media and variational inequalities - In: Free Boundary Problems -
Proc. of a Seminar held in Pavia, Sept.-Oct., 1979, Vol 1, (1980),
175-191.

[14] C. Baiocchi: Disequazioni variazionali, Boll. U.M.I., 18-A(1981),
      173-187.

[15] C. Baiocchi, A Capelo: Disequazioni Variazionali e Quasivariazionali,
      Vol. 1 and 2, Pitagora Editrice (1978) Bologna.

[16] C. Baiocchi, V. Comincioli, E. Magenes and G. A. Pozzi: Free bound-
      ary problems in the theory of fluid flow through porous media:
      Existence and uniqueness theorem. Am. Mat. Pura Appl. 97 (1973),
      1-82.

[17] C. Baiocchi and A. Friedman: A filtration problem in a porous medium
      with variable permeability. Ann. Mat. Pura Appl. 114 (1977), 377-
      393.

[18] A. Bensoussan and J. L. Lions: Nouvelle formulation de problèmes
      de contrôle impulsionnel et application, C. R. Acad. Sc. Paris,
      Série A, 276 (1973).

[19] A Bensoussan and J. L. Lions: Application des Inequations Varia-
      tionnelles en Contrôle Stochastique, Dunod, Paris (1978).

[20] J. Bergh and J. Löfström: Interpolation Spaces: An Introduction.
      Springer (1976).

[21] M. F. Bidaut-Veron: Variational inequalities of order 2m in un-
      bounded domains, Nonlinear Anal. Th. Meth. Appl. 6 (1982), 253-269.

[22] M. Biroli: A De Giorgi-Nash-Moser result for a variational inequality,
      Boll. U.M.I. 16-A (1979), 598-605.

[23] L. Boccardo: Régularité $W_0^{1,p}$ (2 < p < +∞) de la solution d'un problème
      unilatéral. Ann. Fac. Sc. Toulouse 3 (1981), 69-74.

[24] L. Boccardo and F. Murat: Nouveaux resultats de convergence dans
      des problèmes unilatéraux - Nonlinear Partial Differential Equations
      and Their Applications, College de France Seminar, Vol. II, Research
      Notes in Math., Pitman, London (1982) - H. Brezis and J. L. Lions,
      Ed.

[25] J. M. Bony: Principe du maximum dans les espaces de Sobolev, C. R.
      Acad. Sc. Paris, 265 (1967), 333-336.

[26] H. Brezis: Operateurs maximaux monotones et semigroupes de contrac-
      tions dans les espaces de Hilbert, Math. Studies 5, North Holland
      (1975).

[27] H. Brezis: Equations et inéquations nonlineaires dans les espaces
      vectoriels en dualité, Am. Inst. Fourier 18 (1968), 115-175.

[28] H. Brezis: Problèmes unilatéraux, J. Math. Pures Appl. 51 (1972),
      1-168.

[29] H. Brezis: Nouveaux Théorèmes de Régularité pour les problèmes Uni-
      latéraux Rencontre entre Physicien Théoriciens et Mathématicien,
      Strasbourg, Vol. 12 (1971).

[30] H. Brezis: Personal communication.

[31] H. Brezis: Remarque sur l'article de F. Murat, J. Math. Pures Appl.
      60 (1981), 321-322.

[32] H. Brezis, The dam problem revisited. Proc. Sem. Montecatini,
      Pitman, London (1983).

[33] H. Brezis, L. C. Evans, A variational inequality approach to the
      Bellman-Dirichlet equation for two elliptic operators, Arch. Rat.
      Mech. Anal. 71 (1979), 1-13.

[34] H. Brezis and D. Kinderlehrer: The smoothness of solutions to nonlinear variational inequalities, Indiana Univ. Math. J. 23 (1974), 831-844.

[35] H. Brezis, D. Kinderlehrer and S. Stampacchia: Sur une nouvelle formulation du problème de d'écoulement a travers une digue, C. R. Acad. Sc., Paris, Série A 287 (1978), 711-714.

[36] H. Brezis and M. Sibony: Equivalence de deux inequations variationelles et applications. Arch. Rat. Mech. Anal. 41 (1971), 254-265.

[37] H. Brezis and G. Stampacchia: Sur la régularite de la solution d'inéquations elliptiques, Bull. Soc. Math., France 96 (1968), 152-180.

[38] H. Brezis and G. Stampacchia: Remarks on some fourth order variational inequalities, Ann. Scuola Norm. Sup. Pisa 4 (1977), 363-371.

[39] P. L. Butzner and H. Berens: Semigroup of Operators and Approximation, Springer (1967).

[40] L. A. Caffarelli: The regularity of free boundaries in higher dimension, Acta Math. 139 (1978), 155-184.

[41] L. A. Caffarelli: Further regularity in the signorini problem, Comm. in P.D.E., 4 (1979), 1067-1076.

[42] L. A. Caffarelli and A. Friedman: The free boundary for elastic-plastic torsion problems, Trans. Amer. Math. Soc. 252 (1979), 65-97.

[43] L. A. Caffarelli and A. Friedman: Reinforcement problems in elasto-plasticity, Rocky Mount. J. of Math. 10 (1980), 155-184.

[44] L. A. Caffarelli and A. Friedman: The obstacle problem for the biharmonic operator. Ann. Scuola Norm. Sup. Pisa 6 (1979), 151-184.

[45] L. A. Caffarelli and G. Gilardi: Monotonicity of the free boundary in the two dimensional dam problem, Anal. Scol. Norm. Sup., Pisa, 7 (1980), 523-537.

[46] L. A. Caffarelli and D. Kinderlehrer: Potential methods in variational inequalities, Journal d'Anal. Math. 37 (1980), 285-295.

[47] L. A. Caffarelli and N. Riviere: Smoothness and analyticity of free boundaries in variational inequalities. Ann. Scuola Norm. Sup. Pisa 3 (1976), 289-310.

[48] J. Carrillo-Menendez and M. Chipot: Sur l'unicité de la solution du problème de l'écoulement a travers une digue, C. R. Acad. Sc., Paris, Série A, 292 (1981) 191-194.

[49] J. Carrillo-Menendez and M. Chipot: On the dam problem, J. Diff. Eqns. 45 (1982), 234-271.

[50] M. Chipot: Sur la régularité Lipschitzienne de la solution d'inéquations elliptiques, J. Math. Pures Appl. 57 (1978), 69-76.

[51] M. Chipot: Sur la régularité de la solution d'inéquations variationelles elliptiques, C. R. Acad. Sc., Paris, Série A 288 (1979), 543-546.

[52] M. Chipot: On the two obstacle problem in: Free Boundary Problems, Proc. of a Seminar held in Pavia, Sept.-Oct. 1979, Vol. 2, Roma (1980).

[53] M. Chipot: Some Results about an elastic-plastic torsion problem, Nonlinear Anal., Th. Meth. Appl., 3 (1979), 261-270.

[54] M. Chipot: Sur quelques inéquations variationnelles, problème de
      l'écoulement a travers une digue, Thèse d'Etat, Université de Paris
      VI (1981).

[55] R. Courant and D. Hilbert: Methods of Mathematical Physics, Inter-
      science.

[56] R. DeVore and K. Scherer: Interpolation of linear operator on Sobo-
      lev Spaces, Ann. of Math., 109 (1979), 583-599.

[57] G. Duvaut and J. L. Lions: Les Inéquations en Mecanique et en Physi-
      que, Dunod, Paris (1972).

[58] G. Fichera: Probemi elastostatici con vincoli unilaterali: il
      problema di Signorini con ambigue condizioni al contorno. Atti Acad.
      Naz. Lincei Mem., 8 (1963-64), 91-140.

[59] J. Frehse: On the regularity of the solution of a second order
      variational inequality, Boll. Un. Mat. Ital., 6 (1972), 312-315.

[60] J. Frehse: On Signorini's problem and variational problems with
      thin obstacles, Ann. Scuola Norm. Sup., Pisa 4 (1977), 343-362.

[61] A. Friedman: Partial Differential Equations of Parabolic Type,
      Prentice Hall (1964), 196-197.

[62] A. Friedman and R. Jensen: Convexity of the free boundary in the
      Stefan problem and in the dam problem, Arch. Rat. Mech. Anal. 67
      (1978), 1-24.

[63] A. Friedman and G. A. Pozzi: The free boundary for elastic-plastic
      torsion problem, Trans. Amer. Math. Soc. 257 (1980), 411-425.

[64] M. Giaquinta and G. Modica: Regolarità Lipschitziana per la solu-
      zione di alcuni problemi di minimo con vincolo, Ann. Matematica,
      106 (1975), 95-117.

[65] D. Gilbarg and N. S. Trudinger: Elliptic Partial Differential Equa-
      tions of Second Order, Springer (1977).

[66] C. Gerhardt: Regularity of solutions of nonlinear variational
      inequalities, Arch. Rat. Mech. Anal., 52 (1973), 389-393.

[67] J. K. Hale: Ordinary Differential Equations, Wiley (1969), Revised
      edition, Krieger (1980).

[68] P. Hartman and G. Stampacchia: On some nonlinear elliptic differ-
      ential functional equations, Acta Math., 115 (1966), 153-188.

[69] R. Jensen: Boundary regularity for variational inequalities,
      Indiana Univ. Math. J., 29 (1980), 495-504.

[70] D. Kinderlehrer: Variational inequalities with lower dimensional
      obstacles, Israel J. Math., 10 (1971), 339-348.

[71] D. Kinderlehrer: Variational inequalities and free boundary prob-
      lems, Bull. Amer. Math. Soc., 84 (1978), 7-26.

[72] D. Kinderlehrer: The coincidence set of solutions of certain
      variational inequalities, Arch. Rat. Mech. Anal., 40 (1971), 231-
      250.

[73] D. Kinderlehrer: The smoothness of the solution of the boundary
      obstacle problem, J. Math. Pures Appl., 60 (1981), 193-212.

[74] D. Kinderlehrer, L. Nirenberg and J. Spruck: Regularity in elliptic
      free boundary problems, I. J. Anal. Math., 34 (1978), 86-118.

[75] D. Kinderlehrer, L. Nirenberg and J. Spruck: Regularity in elliptic free boundary problems, II. Ann. Scuola Norm. Sup., Pisa, 6 (1979), 637-683.

[76] D. Kinderlehrer and G. Stampacchia: An Introduction to Variational Inequalities and their Applications, Academic Press, 1980.

[77] H. Lanchon: Torsion élastoplastique d'un arbre cylindrique de section simplement on multiplement connexe, Thèse Université de Paris VI (1972).

[78] H. Lanchon: Torsion élastoplastique d'un arbre cylindrique de section simplement on multiplement connexe, Journal de Mecanique, 13 (1974), 267-320.

[79] L. D. Landau and E. M. Lifshitz: Theory of Elasticity, Pergamon Press 1959.

[80] H. Lewy and G. Stampacchia: On the regularity of the solution of a variational inequality, Comm. Pure Appl. Math., 22 (1969), 153-188.

[81] H. Lewy and G. Stampacchia: On the smoothness of superharmonics which solve a minimum problem, J. Anal. Math., 23 (1970), 224-236.

[82] J. L. Lions: Problèmes aux limites dans les équations aux dérivées partielles, Presses Université de Montreal (1962).

[83] J. L. Lions: Quelques methodes de resolution des problemes aux limites nonlineaires, Dunod-Gauthier-Villars, Paris (1969).

[84] J. L. Lions: Sur quelques questions d'analyse de mecanique et de controle optimal, Université de Montreal (1976).

[85] J. L. Lions and E. Magenes: Problèmes aux limites non Homogènes, Vol. I, II, III, Dunod, Paris (1968-70).

[86] J. L. Lions and G. Stampacchia: Variational inequalities. Comm. Pure Appl. Math., 20 (1967), 493-519.

[87] P. L. Lions: Problemes elliptiques du $2^{\text{ème}}$ ordre non sous forme divergence, Proc. Roy. Soc. Edinburgh, 84 (1979), 263-272.

[88] O. Mancino and G. Stampacchia: Convex programming and variational inequalities, J. Opt. Theory Appl., 9 (1972), 3-23.

[89] F. Mignot and J. P. Puel: Inequations d'évolution paraboliques avec convexes dependant du temps. applications aux inéquations quasi-variationnelles d'évolution, Arch. Rat. Mech. Anal., 64 (1977), 59-91.

[90] J. J. Moreau: Proximité et dualité dans un espace hilbertien, Bull. Soc. Math. France, 93 (1965), 273-299.

[91] U. Mosco: Implicit variational problems and quasi variational inequalities, Summer School, Bruxelles 1975, in Lecture Notes in Math. No. 543.

[92] U. Mosco and G. M. Troianiello: On the smoothness of solutions of unilateral Dirichlet problems, Boll. U.M.I., 8 (1973), 57-67.

[93] F. Murat: L'injection du cone postif de H^{-1} dans $W^{-1,q}$ est compacte pour tout $q < 2$. J. Math. Pures Appl., 60 (1981), 309-321.

[94] M. K. Murthy and G. Stampacchia: A variational inequality with mixed boundary conditions, Israel J. Math., 13 (1972), 188-224.

[95] Necas: Les Méthodes directes en théorie des équations elliptiques, Masson (1967).

[96]   M. H. Protter and H. F. Weinberger: A Maximum principle and gradi-
       ent Bounds for linear elliptic equations, Indiana U. Math. J.,
       23 (1973), 239-249.

[97]   M. Schechter: On L^p estimates and regularity I, Amer. J. Math.,
       85 (1963) 1-13; II, Math. Scand., 13 (1963), 47-69.

[98]   L. Schwartz: Théorie des distributions. Hermann, Paris (1966).

[99]   G. Stampacchia: Formes bilinearies coercitives sur les ensembles
       convexes, C. R. Acad. Sc., Paris, 258 (1964), 4413-4416.

[100] G. Stampacchia: Equations Elliptiques du Second Ordre à Coeffici-
       ents Discontinus, Presse de l'Universite de Montreal (1965).

[101] G. Stampacchia: Regularity of solutions of some variational in-
       equalities, Nonlinear functional analysis (Proceedings Symp. Pure
       Math., Vol. 18, Part I, 271-281).

[102] G. Stampacchia: On the filtration of a liquid through a porous
       medium, Usp. Mat. Nauk., 29 (1974), 89-101.

[103] E. M. Stein: Singular Integrals and Differentialibility Properties
       of Functions, Princeton University Press (1970).

[104] L. Tartar: Interpolation nonlineaire et régularité. J. Funct.
       Anal., 9 (1972), 469-489.

[105] L. Tartar: Inéquations quasivariationnelles abstraites, C. R.
       Acad. Sc., Paris, Série A, 278 (1974), 1193-1196.

[106] T. W. Ting: Elastic-plastic torsion of simply connected cylindrical
       bars, Indiana Univ. Math. J., 20 (1971), 1047-1076.

[107] T. W. Ting: Elastic-plastic torsion problem over multiple connected
       domains, Ann. Scuola Norm. Sup., Pisa, 4 (1977), 291-312.

[108] H. Triebel: Interpolation Theory, Function Spaces, Differential
       Operators, North Holland (1978).

[109] A. Visintin: Study of a free boundary filtration problem by a
       nonlinear variational equation, Boll. U.M.I., 5 (1979), 212-237.

Index

Baiocchi transform, 94

Bellman-Dirichlet problem, 32

Bilinear form, 8

Chain rule for piecewise smooth functions, 13

Coercive operator, 4

Coincidence set, 31

Compact embedding, 13

Continuous on finite dimensional subspaces, 5

Convex function, 3

Convex set, 3

Dam problem, 74-110
  existence of solutions, 79, 81
  monotonicity of free boundary, 106
  nonuniqueness of solutions,
  properties of solutions, 82-90
  strong formulation, 76
  uniqueness of S_3-connected solutions, 101
  weak formulation, 78

Darcy's law, 76

Dirichlet problem, (see elliptic equation)

Elastic-plastic torsion problem, 17, 52

Elliptic equation
  existence of solutions, 14, 15
  L^p-estimates, 67-72
  nonlinear, 15

Exterior sphere condition, 53

Fixed point theorem, 1

Fourth order variational inequality, 21

Free boundary, 31, 86, 106, 107

Hilbert space, 8

Hölder continuous functions, 14

Interpolation spaces, 60-62

Laplacian, 14

Lax-Milgram theorem, 8

L^p-spaces, 10

Minty's Lemma, 6

Monotone operator, 5
  strictly, 5
  surjectivity of, 7

Negative part of a function, 13

Norms, 10

Numerical analysis, 21

Obstacle problems, 16, 18, 22-66

    boundary estimates, 35

    comparison principle, 22

    continuity of solutions, 30

    existence of solutions, 16, 18

    free boundary, 31

    $W^{1,p}$ regularity, 60-66

    $W^{1,\infty}$ regularity, 50-60

    $W^{2,p}$ regularity, 27-30

    $W^{2,\infty}$ regularity, 33-50

Parabolic variational inequalities, 21

Penalization, 79

Penalty method, 24-26, 73

Permeability, 76

Poincaré's inequality, 11

Pool, 92

Positive part of a function, 13

Porous medium, 74

Prescribed mean curvature operator, 20

Projection on a convex set, 3

Quasi-variational inequality, 9

Regularity, (see obstacle problems)

Reservoir, 75

Schauder fixed point theorem, 1

Sobolev embedding theorems, 12, 13

Sobolev spaces, 10-13, 21

Spaces of continuous functions, 14

Variational inequality, 2, (see also dam problem, elastic-plastic torsion problem, and obstacle problems)

    existence of solutions, 3, 4, 6

    fourth order, 20

    parabolic, 21

    uniqueness of solutions, 7

Wet set, 76, 84, 85